Adaptive Array Systems

Adaptive Array Systems

Fundamentals and Applications

B. Allen and M. Ghavami

Both of
Centre for Telecommunications Research
King's College London, UK

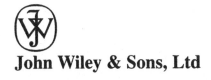

John Wiley & Sons, Ltd

Other Wiley Editorial Offices

John Wiley & Sons Inc., 111 River Street, Hoboken, NJ 07030, USA

Jossey-Bass, 989 Market Street, San Francisco, CA 94103-1741, USA

Wiley-VCH Verlag GmbH, Boschstr. 12, D-69469 Weinheim, Germany

John Wiley & Sons Australia Ltd, 33 Park Road, Milton, Queensland 4064, Australia

John Wiley & Sons (Asia) Pte Ltd, 2 Clementi Loop #02-01, Jin Xing Distripark,
Singapore 129809

John Wiley & Sons Canada Ltd, 22 Worcester Road, Etobicoke, Ontario,
Canada M9W 1L1

Wiley also publishes its books in a variety of electronic formats. Some content that
appears in print may not be available in electronic books.

British Library Cataloguing in Publication Data

A catalogue record for this book is available from the British Library

ISBN 0-470-86189-4

Typeset by the authors using LATEX software.

Contents

Preface

Readership

Firstly, this book is set at a level suitable for senior undergraduate and postgraduate students who wish to understand the fundamentals and applications of adaptive array antenna systems. Array fundamentals are described in the text, and examples which demonstrate theoretical concepts are included throughout the book, as well as summaries and questions at the end of each chapter. We also consider this book to be useful for researchers, practising engineers and managers alike, especially where an accessible text on adaptive array fundamentals and applications is required. A complete chapter on implementation aspects highlights the challenges a designer may encounter during the development of an array system. The book contains explanations of fundamentals, description of algorithms and presentation of research-based case studies making it appeal to a wide readership.

About the Book

This book aims to provide an accessible text on adaptive array fundamentals. Although the book considers a wide range of applications, including wireless communications, radar, sonar and bio-medical, the focus is predominantly wireless communications. This reflects the research interests of the authors, but it should be noted that many of the techniques introduced throughout the text can be applied to other branches of engineering such as bio-medical.

The book is structured as follows. Chapter 1 (Fundamentals of Array Signal Processing) introduces antenna and sensor terminology and then discusses common antenna elements and reviews the characteristics of each. The chapter concludes by reviewing the array system concept. In chapter 2 (Narrowband Array Systems), the behaviour of narrowband antenna arrays is analysed. The function of phase and amplitude weights and beamsteering is explored within the context of a narrowband array, and the phenomenon of grating lobes is introduced. The chapter concludes by characterising a number of common window functions that are used to control the sidelobe levels of array beam-patterns. Chapter 3 (Wideband Array Processing) then introduces and analyses several wideband beamforming techniques and compares the performance of each. The focus of chapter 4 (Adaptive Arrays) is on algorithms, where a range of adaptive algorithms and direction of arrival algorithms are presented and discussed. The chapter concludes with a review of several blind beamforming algorithms and a comparison of direction of arrival estimation performance. Chapter 5 (Practical Considerations) contains a wide variety of topics that relate to implementation aspects of adaptive arrays. These include hardware implementation aspects, circular arrays, channel modelling and transmit beamforming. Finally, chapter 6 (Applications) discusses the application of adaptive array systems through several detailed case studies on:

- wideband arrays, radar, sonar and bio-medical imaging;

- second- and third-generation terrestrial wireless systems; and

- satellite communication systems.

Several texts already exist on adaptive arrays as it is a topic that has undergone significant research and development over the last 50 years. These texts can be broadly classified as follows:

- rigorous mathematical treatment;

- emphasis on radar; and

- emphasis on wireless communications.

In contrast to many of these texts, we have tried to make this book readable and accessible to the uninitiated, with a broad range of applications considered and full chapters covering wideband beamforming and implementation aspects. We have also included research-based material making it appeal to the experienced researcher as well.

Some consider adaptive antennas to be a mature technology with little research left to tackle. Contrary to this belief, we consider there to be an ongoing interest in adaptive antennas for future (3G and 4G) mobile communications systems, ultra wideband (UWB) wireless systems where the signal bandwidth is very large, and satellite communication and navigation systems. In particular, UWB presents many research challenges in the area of communications and bio-medical engineering and the applications of antenna arrays can provide performance gains to both.

We consider this text to be unique because it covers array fundamentals for a wide range of applications, as well as specifically covering implementation aspects and applications through detailed case studies. Examples are included throughout the text which illustrate the concepts under discussion and we have attempted to write the text in an accessible and appealing way.

Prerequisites

In order for the most to be gained from the contents of this book, it is recommended (but not essential) that the reader has a firm grounding in the principles of:

- engineering mathematics, including Fourier analysis and matrix algebra;

- signals and systems;

- electromagnetics;

- radiowave propagation;

- radio frequency circuit design; and

- communications engineering principles.

Course Design

This book has been designed in such a way that it forms a complete semester's course on adaptive array systems. We suggest that such a self-contained course consists of four-hours of lectures and a two-hour tutorial for each chapter, with actual times being adjusted according to ability. In particular, chapters 1 and 2 present the fundamentals of radiation, antenna elements and beamforming suitable for taught courses, and chapter 4 contains fundamental signal processing concepts for adaptive arrays. These topics can be complemented with implementation issues and case studies in chapters 5 and 6. In contrast, chapter 3 is particularly suited for the research-active readership, as is substantial sections of chapters 5 and 6 where novel developments are reported, especially with regard to wideband beamforming algorithms and channel modelling.

As an extra resource, the companion website for our book contains a solutions manual, Matlab m-files for the examples and problems, and a sample chapter.

Also, for those wishing to use this material for lecturing purposes, electronic versions of some of the figures are available. Please go to the following URL and take a look: ftp://ftp.wiley.co.uk/pub/books/allen.

We hope that you find this book useful both as a reference, a learning tool and a stepping stone to further your own efforts in this multi-disciplinary field of engineering.

B. Allen

M. Ghavami

London

Acknowledgments

Our thanks are extended to our colleagues at the Centre for Telecommunications Research, King's College London for providing a rich research environment that has enabled the timely development of this book. In particular, we wish to thank Professor Hamid Aghvami for his leadership and provision of opportunity. Also, we especially thank Dr Mischa Dohler, Neville Tyler, Adil Shah and Dora Karveli for contributing to this book. Without these contributions the book would not have reached its current form. Thank you!

The authors wish to express their gratitude to Professors Joe McGeehan, Andy Nix, Mark Beach, and Dr Geoff Hilton of the University of Bristol, who have conducted substantial research into adaptive antenna systems that has aided in the formulation of this book. In particular, we wish to thank Professor Mark Beach for reviewing this book and Ben wishes to thank him for his expert PhD supervision that has provided invaluable inspiration.

Our gratitude goes to the IEEE and IEICE for granting permission to incorporate substantial sections of published works in chapters 3 and 6. We would also like to thank Sarah Hinton at Wiley for assisting with the organisation, marketing and production of this book, and to the anonymous copy editor for the attention to detail that was evident in the corrections.

Last, but certainly not least, Ben wishes to thank:

his wife, Louisa, for her love and companionship during the last few years, during which a wide variety of challenges have been faced by us. He also wishes to dedicate this book to Louisa and his ancestors: Leslie (Bill) Allen and William Smith who spent their working lives involved with early telecommunications technologies.

and Mohammad wishes to thank:

his wife, Mahnaz, and children for their love and patience during the period of preparation for this book. Mohammad would like to dedicate this book in memory of his father, Reza.

List of Figures

List of Tables

Introduction

I.1 ADAPTIVE FILTERING

In recent years *adaptive signal processing* has been applied to a variety of areas, ranging from bio-medical applications to telecommunications networks. Such filters can self-adjust to the characteristics of an incoming signal without external intervention. This is similar to a closed-loop control system such as a thermostatically controlled central heating system. Well-known applications of adaptive signal processing include:

- cancelling the maternal heart beat when performing fetal electrocardiogra‑ ‧phy;

- noise cancellation in speech signals;

- echo cancellation in long distance telephone circuits; and

- adaptive equalisation in mobile communication systems.

These filters are all based upon time (or frequency) domain filtering, often utilising familiar *finite impulse response* (FIR) or *infinite impulse response* (IIR) filter architectures, where the filter taps are computed according to some criteria and the incoming signal characteristics.

Over time a number of applications have been developed where *spatial filtering* is desirable. The spatial filtering principle is demonstrated by considering the alignment of a TV antenna, where misalignment either results in no signal, or reception of a distant unwanted transmitter. Also, careful choice and alignment of the antenna can reduce the effect of *ghosting* which is caused by receiving the wanted signal and an echo reflected from a nearby structure (i.e., multipath). Conversely, correct alignment results in good, echo-free reception. Thus, spatial filtering can reject interfering signals, eliminate multipath signals and enhance the wanted signal.

Instead of the manual alignment of the antenna, spatial filtering can be achieved electronically using signal processing techniques derived from the time domain adaptive signal processing techniques mentioned previously. Such a system consists of a number of sensors, followed by signal processing which can be implemented using either analogue or digital technologies, or a hybrid. Depending upon the application, the sensors could be:

- antenna elements for electromagnetic signals (i.e., radar, radio);

- hydrophones for acoustic signals (i.e., sonar);

- seismometers for seismic signals; or

- microphones for cardiac or other audio signals.

These multi-technology systems are often referred to as: *beamformers*; *adaptive arrays* (if they are adaptive), *smart antennas* (for wireless systems); *phased arrays* (often used in the context of radar); or *space-time processors* (a generic term).

I.2 HISTORICAL ASPECTS

An early simple adaptive antenna was the *adaptive sidelobe canceller*, which was developed in the late 1950s by P. Howells [1] and subsequently by P. Howells and S. Applebaum [2]. Such a system is shown in figure I.1, where two omni-directional antennas are used in a similar manner to microphones in an adaptive noise canceller, i.e., one to provide a reference signal and the other to provide an input to the adaptive filter. Both antennas receive the interfering and wanted signals, but since the antennas are spatially separated the two signals can be distinguished using a suitable filter. This configuration works very well as long as the input *signal-to-interference ratio* (SIR) is low. Indeed, when the SIR is high, the wanted signal can be attenuated! This is because the system directs a null towards a specific signal and under these conditions the null is directed towards the wanted signal. There are several techniques that can reduce this effect such as injecting additional noise, or using a specially modified adaptive algorithm.

Another type of adaptive beamformer utilises a pilot signal. This technique was developed by Widrow, Mantey, Griffiths and Goode [3]. It operates by forming a beam towards the wanted signal whilst simultaneously directing nulls

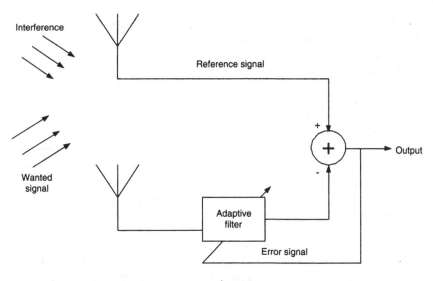

Fig. I.1 Two-element adaptive sidelobe canceller.

towards interfering signals. This differs from the adaptive sidelobe canceller which can only steer one null. Griffiths [4] and Frost [5] have also developed similar systems which have been shown to be simpler and, in some cases, perform better. In 1976, Swarner, Ksienski Compton and Huff [6] applied the pilot adaptive beamformer to wireless communication systems, where prior to this radar was the primary application. This new application was key to an overwhelming amount of subsequent research, a cross-section of which is reported in this book.

I.3   CONCEPT OF SPATIAL SIGNAL PROCESSING

Having two ears gives us the ability to sense the direction of audio signals and to adjust our hearing to 'tune' into certain signals and ignore others. Moreover, human hearing covers a frequency range of approximately 5 Hz to 20 kHz. Due to the bandwidth, designing such a system electronically is highly challenging and requires specialised wideband array design techniques that account for the signal characteristics as they change over the operating bandwidth.

The remarkable ability of human hearing to adapt to certain signals forms the basis of a wideband adaptive array. Such an array can be designed for audio signals or electromagnetic signals and therefore has applications in sonar, radar, and any other system that can benefit from the spatial processing of wideband signals.

One such area that has blossomed recently is multimedia communications where bandwidth limitations have meant that advanced techniques such as adaptive antenna systems are required to provide the necessary channel capacity [7].

Traditionally, wireless signals have occupied bandwidths of up to 20 MHz. However, ultra-wideband wireless systems, which occupy several gigahertz, have recently become a candidate for delivery of unprecedented data rates over short ranges [8], and this has provided a demand for beamformers to operate over these bandwidths.

Adaptive arrays can also be applied to radar and sonar systems that enable the direction of targets to be determined. One key advantage adaptive antennas have is that they eliminate the need to mechanically rotate or move the sensors. However, rotating radar antennas are still a common sight on ships and at airports.

The study of adaptive antenna arrays is multi-disciplinary, demanding a knowledge of signal processing, transceiver design, antenna design and propagation. Without any one of these component disciplines, it becomes very difficult to design and commission an effective and robust system. As a consequence, it has attracted the attention of many of the world's top researchers and enormous investment from a broad range of commercial and government organisations. This research has been supported by the rapid growth of the electronics industry over the last few decades. As a consequence, cost-effective components, such as analogue-to-digital converters and signal processing technology. Also, the opening up of the mobile communications marketplace has been a key driver for adaptive antenna development.

Smart antennas have been considered as a key technology to provide the services demanded by mobile network users. This is due to their ability to control interference levels, thereby allowing additional services to be provided without causing unacceptable additional interference levels. Although smart antennas have not been widely integrated into cellular systems as many have predicted, there have been many examples of their use, particularly in rural areas where they are used to increase base stations coverage. Also, a possible future application of smart antennas is as an enabling technology for executing spectrum liberalisation. Spectrum liberalisation, which is currently being considered by spectrum planning authorities, is the de-regulating of the civil radio spectrum to allow users to operate on any frequency (within certain constraints) assuming the interference levels are of an acceptable level. This is in contrast to existing policy where users are assigned a specific operating frequency. In this context, smart antennas will control interference levels received and generated by the transceiver, and would operate in conjunction with software definable radios and associated protocols.

1

Fundamentals of Array Signal Processing

1.1 INTRODUCTION

The robust design of an adaptive array system is a multi-disciplinary process, where component technologies include: signal processing, transceiver design, array design, antenna element design, and signal propagation characteristics. The 'glue' between the technologies is provided by the system engineer who specifies the requirements for each, so that the complete adaptive array system will operate according to the required performance criteria.

This chapter introduces the fundamentals that enable the design of these component technologies and sets the scene for much of the remainder of the book. It is an important prerequisite for chapters 2–5 where the fundamentals are expended to include an understanding of adaptive antenna arrays as well as practical issues related to adaptive antenna design.

The first section of this chapter is dedicated to the very fundamentals of antennas, where the reasons why antennas are capable of transmitting and receiving an electromagnetic signal are explained. Surprisingly, transmission and reception are facilitated by a small asymmetry in Maxwell's equations, which triggers an electromagnetic wave to decouple from (transmission) and couple into (reception) a medium carrying free electric charges (antenna). The application of Maxwell's equations to the most fundamental radiation element, the Hertzian dipole, is then explained. It allows some common antenna terminology to be defined, which is not confined to Hertzian dipoles only but is also applicable to practical antenna elements.

Adaptive Array Systems   B. Allen and M. Ghavami
© 2005 John Wiley & Sons, Ltd   ISBN 0-470-86189-4

This brings us to a brief description of typically occurring antenna elements, which themselves might be part of larger antenna arrays which are subsequently introduced. Here, commonly occurring antenna arrays, such as the linear and circular antenna arrays, are discussed. Antenna arrays are then applied to achieve spatial filtering which enhances the signal strength (beamforming) and weakens the interference power (nulling) in wireless communication systems.

A favourable candidate to accomplish spatial filtering is the adaptive antenna array, where channel parameters are fed into the algorithms controlling the adaptive beamforming so as to optimise the performance. The conditions under which beamforming can be achieved and which inter-element spacings are required are then explained.

The chapter is finalised with the conclusions and some problems related to the fundamentals of antenna arrays.

1.2   THE KEY TO TRANSMISSION

1.2.1   Maxwell's Equations

Starting from some assumptions and observations, Maxwell derived a set of mutually coupled equations, the so called *Maxwell Equations*, which paved the way to the field of electrodynamics, part of which allows a proper understanding and design of antenna elements and arrays. In differential form, the four equations are:

$$\text{div } \mathbf{D}(\mathbf{r}, t) = \rho(\mathbf{r}, t) \tag{1.1}$$

$$\text{div } \mathbf{B}(\mathbf{r}, t) = 0 \tag{1.2}$$

$$\text{curl } \mathbf{E}(\mathbf{r}, t) = -\frac{\partial \mathbf{B}(\mathbf{r}, t)}{\partial t} \tag{1.3}$$

$$\text{curl } \mathbf{H}(\mathbf{r}, t) = \frac{\partial \mathbf{D}(\mathbf{r}, t)}{\partial t} + \mathbf{J}(\mathbf{r}, t) \tag{1.4}$$

where $\mathbf{E}(\mathbf{r}, t)$ in [V/m] is the vector representing the *electric field intensity*, $\mathbf{D}(\mathbf{r}, t)$ in [C/m^2] is the *electric flux density*, $\mathbf{H}(\mathbf{r}, t)$ in [A/m] is the *magnetic field intensity*, $\mathbf{B}(\mathbf{r}, t)$ in [T] is the *magnetic flux density*, $\rho(\mathbf{r}, t)$ in [C/m^3] is the *charge density* and $\mathbf{J}(\mathbf{r}, t)$ in [A/m^2] is the *current density*. All of the above electromagnetic field variables depend on the spatial position with respect to some coordinate system, $\mathbf{r}$ in [m], and the elapsed time, t in [s].

The electric and magnetic field vectors can be related through the material equations, i.e.

$$\mathbf{D}(\mathbf{r}, t) = \epsilon_0 \epsilon_r \mathbf{E}(\mathbf{r}, t) \tag{1.5}$$

$$\mathbf{B}(\mathbf{r}, t) = \mu_0 \mu_r \mathbf{H}(\mathbf{r}, t) \tag{1.6}$$

where $\epsilon_0 \approx 8.85 \cdot 10^{-12}$ [F/m] is the free space permittivity, ϵ_r is the material dependent *relative permittivity* (also called the *dielectric constant*), $\mu_0 \approx 1.257 \cdot 10^{-6}$ [H/m] is the *free space permeability* and μ_r is the material dependent *relative permeability*.

Finally, the *div* operation characterises how much a vector field linearly diverges and the *curl* operation characterises the strength of the curl (rotation) in the field. Both relate to spatial operations, i.e. they do not involve any operations with respect to time.

1.2.2 Interpretation

To understand the meaning of the mathematical formulation of the above equations, let's scrutinise them one by one.

$\operatorname{div} \mathbf{D}(\mathbf{r}, t) = \rho(\mathbf{r}, t)$ can be rewritten as $\rho(\mathbf{r}, t) = \operatorname{div} \mathbf{D}(\mathbf{r}, t)$, i.e. static or dynamic charges in a given volume are responsible for a diverging electric field. This implies that there must be a distinct source and sink for the electric field since a field cannot possibly (linearly) diverge and start and end in the same location.

$\operatorname{div} \mathbf{B}(\mathbf{r}, t) = 0$ can be rewritten as $0 = \operatorname{div} \mathbf{B}(\mathbf{r}, t)$, i.e., there is nothing available in nature which makes a magnetic field diverge. This equation comes from the observation that there are no magnetic 'charges' known to physics. Note that magnetic charges are sometimes introduced in theoretical electrodynamics so as to simplify the derivation of certain theories.

$\operatorname{curl} \mathbf{E}(\mathbf{r}, t) = -\partial \mathbf{B}(\mathbf{r}, t)/\partial t$ means that a spatially varying (curling) electric field will cause a time-varying magnetic field. Alternatively, it can be rewritten as $-\partial \mathbf{B}(\mathbf{r}, t)/\partial t = \operatorname{curl} \mathbf{E}(\mathbf{r}, t)$, i.e. a time-varying electric field will cause a curl in the magnetic field.

Finally, and most importantly as shown shortly, $\operatorname{curl} \mathbf{H}(\mathbf{r}, t) = \partial \mathbf{D}(\mathbf{r}, t)/\partial t + \mathbf{J}(\mathbf{r}, t)$ can be read as follows. A spatially varying (curling) magnetic field will cause a time-varying electric field <u>and</u>, if existent, also a current through a medium capable of carrying a flow of electric charges. The equation can also be read as either a current flow through a medium or a time-varying electric field producing a spatially curling magnetic field.

1.2.3 Key to Antennas

The first two equations yield separately an insight into the properties of the electric field and magnetic field, respectively. The remaining two equations, however, show that both fields are closely coupled through spatial (curl) and temporal $(\partial/\partial t)$ operations. It can also be observed that the equations are entirely symmetric - apart from the current density $\mathbf{J}(\mathbf{r}, t)$. It turns out that this minor asymmetry is responsible for any radiation process occurring in nature, including the transmission and reception of electromagnetic waves. For the ease of explanation, the last two of Maxwell equations are rewritten as

$$\operatorname{curl} \mathbf{E}(\mathbf{r}, t) = -\mu_0 \mu_r \frac{\partial \mathbf{H}(\mathbf{r}, t)}{\partial t} \tag{1.7}$$

$$\operatorname{curl} \mathbf{H}(\mathbf{r}, t) = \quad \epsilon_0 \epsilon_r \frac{\partial \mathbf{E}(\mathbf{r}, t)}{\partial t} + \mathbf{J}(\mathbf{r}, t) \tag{1.8}$$

Let us assume first that there is a static current density $\mathbf{J}(\mathbf{r})$ available which, according to (1.8), causes a spatially curling magnetic field $\mathbf{H}(\mathbf{r})$; however, it fails to generate a temporally varying magnetic field which means that $\partial \mathbf{H}(\mathbf{r})/\partial t = 0$. According to (1.7), this in turn fails to generate a spatially and temporally varying electric field $\mathbf{E}(\mathbf{r})$. Therefore, a magnetic field is only generated in the location where a current density $\mathbf{J}(\mathbf{r})$ is present. Since the focus is on making a wave propagating in a wire<u>less</u> environment where no charges (and hence current densities) can be supported, a static current density $\mathbf{J}(\mathbf{r})$ is of little use.

The observations, however, change when a time-varying current density $\mathbf{J}(\mathbf{r}, t)$ is generated, which, according to (1.8), generates a spatially and temporally varying magnetic field $\mathbf{H}(\mathbf{r}, t)$. Clearly, $\partial \mathbf{H}(\mathbf{r}, t)/\partial t \neq 0$ which, according to (1.8), generates a spatially and temporally varying electric field $\mathbf{E}(\mathbf{r}, t)$, i.e., $\partial \mathbf{E}(\mathbf{r}, t)/\partial t \neq 0$. With reference to (1.8), this generates a spatially and temporally varying magnetic field $\mathbf{H}(\mathbf{r}, t)$, even in the <u>absence</u> of a current density $\mathbf{J}(\mathbf{r}, t)$, and so on.

A wave is hence born where the electric field stimulates the magnetic field and vice versa. From now on, this wave is referred to as an electromagnetic (EM) wave, since it contains both magnetic and electric fields. From the above it is clear that such a wave can now propagate without the need of a charge-bearing medium; however, such a medium can certainly enhance or weaken the strength of the electromagnetic wave by means of an actively or passively created current density $\mathbf{J}(\mathbf{r}, t)$.

In summary, to make an electromagnetic wave decouple from a transmitting antenna, a medium capable of carrying a time-varying current density $\mathbf{J}(\mathbf{r}, t)$ is required. A medium which achieves this with a high efficiency is called an *antenna*. As simple as that! An antenna can hence be anything: a rod, wire, metallic volumes and surfaces, etc.

Remember that $\mathbf{J}(\mathbf{r}, t) = \partial Q/\partial t$, where Q in [C] is the electric charge; therefore, if a time-varying current density for which $\partial \mathbf{J}(\mathbf{r}, t)/\partial t \neq 0$ is required, it must be ensured that $\partial^2 Q/\partial t^2 \neq 0$, i.e. that the charges are accelerated. Sometimes such acceleration happens unintentionally, e.g. in bent wires where electrons in the outer radius move faster than the ones in the inner radius of the bend. Therefore, any non-straight piece of wire will emit electromagnetic waves which, in the case of antennas, is desirable, but in the case e.g. of wiring between electric components in a computer, not at all.

1.3   HERTZIAN DIPOLE

A wire of infinitesimal small length δl is known as a Hertzian dipole. It plays a fundamental role in the understanding of finite length antenna elements, because any of these consists of an infinite number of Hertzian dipoles. Antenna arrays, on the other hand, can be represented by means of a plurality of finite length antenna elements. This, somehow, justifies the importance of properly understanding the radiation behaviour of a Hertzian dipole.

The Hertzian dipole can be fed by a current $I(t)$ of any temporal characteristics; however, since an arbitrary signal can be resolved into its spectral harmonics, the focus is on the analysis of sinusoidal feeding, i.e. a single frequency. Thus currents of the form $I(t) = I_{\max} \cdot \exp(j\omega t)$ are considered, where $I_{\max}$ is the maximum current and ω in [rad/s] is the angular frequency. Once the response to a particular spectral component has been determined, the total response to an arbitrary temporal excitation is obtained by linearly superimposing the individual spectral contributions.

With these assumptions, it is straightforward to show that a feeding current of $I(t) = I_{\max} \cdot \exp(j\omega t)$ will cause an EM wave of the form $\mathbf{E}(\mathbf{r}, t) = \mathbf{E}(\mathbf{r}) \cdot \exp(j\omega t)$ and $\mathbf{H}(\mathbf{r}, t) = \mathbf{H}(\mathbf{r}) \cdot \exp(j\omega t)$. Focus is given to the spatial dependencies of the field components and henceforth we omit the harmonic temporal factor $\exp(j\omega t)$.

A Hertzian dipole with a feeding generator is depicted in figure 1.1. For obvious reasons, the coordinate system of choice is spherical where a point in space is characterised by the azimuth $\phi \in (1, 2\pi)$, elevation $\theta \in (1, \pi)$ and distance from the origin r. The field vectors can therefore be represented as $\mathbf{E}(\mathbf{r}) = (E_\phi, E_\theta, E_r)$ and $\mathbf{H}(\mathbf{r}) = (H_\phi, H_\theta, H_r)$.

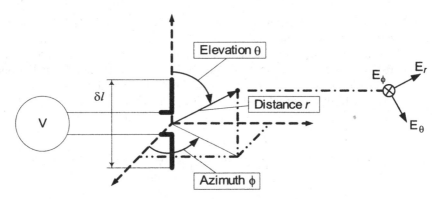

Fig. 1.1   Feeding arrangement and coordinate system for Hertzian dipole.

To obtain these electromagnetic field contributions radiated by a Hertzian dipole, one needs to solve Maxwell's equations (1.1)–(1.4). The procedure, although straightforward, is a bit lengthy and is hence omitted here. The interested

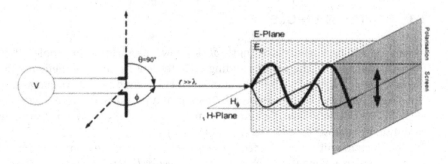

Fig. 1.2   Relationship between **E** and **H** in the far-field.

reader is referred to [9]. The final solution is of interest to us, where the non-zero EM field components are found to be

$$E_\theta = -\frac{\eta k^2}{4\pi} I \delta l \; \sin(\theta) \; e^{-jkr} \left[\frac{1}{jkr} + \left(\frac{1}{jkr}\right)^2 + \left(\frac{1}{jkr}\right)^3 \right] \qquad (1.9)$$

$$E_r = -\frac{\eta k^2}{2\pi} I \delta l \; \cos(\theta) \; e^{-jkr} \left[\left(\frac{1}{jkr}\right)^2 + \left(\frac{1}{jkr}\right)^3 \right] \qquad (1.10)$$

$$H_\phi = -\frac{k^2}{4\pi} I \delta l \; \sin(\theta) \; e^{-jkr} \left[\frac{1}{jkr} + \left(\frac{1}{jkr}\right)^2 \right] \qquad (1.11)$$

and all remaining components are zero, i.e. $E_\phi = 0$ and $H_\theta = H_r = 0$. Here, $k = 2\pi/\lambda$ is the wave number and λ in [m] is the wavelength. These equations can be simplified for distances near to and far from the dipole, where the latter is obviously of more importance to the aspects covered here, i.e., far-field operation.

The far-field, also referred to as the Fraunhofer region, is characterised by $kr \gg 1$, which allows simplification of (1.9)–(1.11) to

$$E_\theta = \eta \, H_\phi \qquad (1.12)$$

$$H_\phi = j\frac{k^2}{4\pi} I \delta l \; \frac{\sin(\theta) \, e^{-jkr}}{kr} \qquad (1.13)$$

and $E_\phi = E_r = 0$ and $H_\theta = H_r = 0$. The wave is visualised in figure 1.2 for $\theta = \pi/2$, i.e. perpendicular to the Hertzian dipole. In the far-field, **E** and **H** are clearly in-phase, i.e. they have maxima and minima at the same locations and times. Also, the EM wave turns into a *plane wave* consisting of two mutually orthogonal electromagnetic field components.

1.4 ANTENNA PARAMETERS & TERMINOLOGY

The brief introduction to the radiation behaviour of an infinitesimal radiating element allows us to introduce some concepts which are vital for the characterisation of antenna elements and arrays. The following introduces important antenna terminology that are commonly used when characterising antennas and arrays.

1.4.1 Polarisation

From (1.12) and figure 1.2, it is observed that in the far-field from the dipole the E-wave oscillates in a plane. It has also been seen that the H-wave oscillates perpendicular to the E-wave. Since both waves always occur together, it is, unless otherwise mentioned, from now on referred to as the E-wave.

If a plane orthogonal to the direction of propagation is cut, indicated as the polarisation screen in figure 1.2, the electric field vector E_θ is observed oscillating on a straight line. This polarisation state is referred to as *linear polarisation* and is further illustrated in figure 1.3.

If two orthogonal dipoles are considered instead of one, and fed with in-phase currents, then this will trigger two decoupled EM waves to be orthogonal in the far-field. A tilted but straight line on the orthogonally cut screen is then observed; this polarisation state is often referred to as *linear tilted polarisation*. If, on the other hand, both dipoles are fed with currents in quadrature phase, i.e. shifted by 90°, then the resulting E-field will be *circularly polarised*. Finally, if the amplitudes of the two decoupled fields are different due to different feeding current amplitudes, then the resulting polarisation will be *elliptical*. In dependency of whether the two feeding currents are +90° or −90° shifted, the polarisation will be left or right circular or elliptical polarised. Finally, if the two feeding currents deviate by a phase different from 90° and/or more Hertzian dipoles are used, then more complicated polarisation patterns can be obtained.

1.4.2 Power Density

The instantaneous power density, w, in $[\text{W/m}^2]$ is defined as

$$w = \text{Re}\left\{ \mathbf{E} \times \mathbf{H}^* \right\} \tag{1.14}$$

where $\text{Re}(x)$ denotes the real part of x, and $\mathbf{H}^*$ is the complex conjugate of $\mathbf{H}$. The average power density can be obtained from (1.14) by assuming that the EM-wave is harmonic, which yields

$$\overline{w} = \text{Re}\left\{ \mathbf{S} \right\} = \text{Re}\left\{ \frac{1}{2}\mathbf{E} \times \mathbf{H}^* \right\} \tag{1.15}$$

where $\mathbf{S}$ is referred to as the Pointing vector.

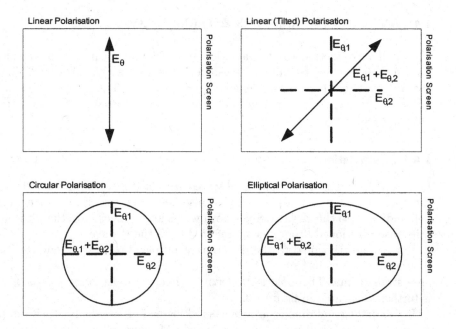

Fig. 1.3   Linear, linear tilted, circular and elliptical polarisation states.

As an example, let us calculate the average power density of an EM-wave in the far-field for a Hertzian dipole, where equations (1.12) and (1.13) are utilised to obtain

$$\mathrm{Re}\,\{\mathbf{S}\} = \mathrm{Re}\left\{ \frac{1}{2} \begin{vmatrix} \mathbf{e}_r & \mathbf{e}_\theta & \mathbf{e}_\phi \\ 0 & E_\theta & 0 \\ 0 & 0 & H_\phi^* \end{vmatrix} \right\} \tag{1.16}$$

$$= \frac{\eta I^2 \sin^2(\theta)}{8r^2} \left(\frac{\delta l}{\lambda} \right)^2 \mathbf{e}_r \tag{1.17}$$

where $\mathbf{e}_r, \mathbf{e}_\theta$ and $\mathbf{e}_\phi$ are the unit vectors of the perpendicular spherical coordinates.

1.4.3   Radiated Power

Given the power density, the total average power passing through a sphere with surface area σ is calculated as

$$P = \int_\sigma \mathrm{Re}\,\{\mathbf{S}\}\, d\sigma \tag{1.18}$$

$$= \int_0^{2\pi} \int_0^\pi \mathrm{Re}\,\{\mathbf{S}\}\, r^2 \sin(\theta)\, d\theta\, d\phi \tag{1.19}$$

which yields for the Hertzian dipole

$$P = \frac{\pi \eta I^2}{3} \left(\frac{\delta l}{\lambda} \right)^2. \tag{1.20}$$

Clearly, the average radiated power is independent of the distance but does depend on the geometrical (δl) and electrical (λ) properties of the dipole.

1.4.4 Radiation Resistance

The radiation resistance, R_r, is defined as *the value of a hypothetical resistor which dissipates a power equal to the power radiated by the antenna when fed by the same current I*, i.e.

$$\frac{1}{2} IU = \frac{1}{2} I^2 R_r = P \tag{1.21}$$

With reference to (1.20), this simply yields for the Hertzian dipole:

$$R_r = \frac{2\pi \eta}{3} \left(\frac{\delta l}{\lambda} \right)^2 = 80\pi^2 \left(\frac{\delta l}{\lambda} \right)^2 = 789 \left(\frac{\delta l}{\lambda} \right)^2 \tag{1.22}$$

the unit of which is $[\Omega]$.

1.4.5 Antenna Impedance

The antenna impedance, Z_a, is defined as *the ratio of the voltage at the feeding point V(0) of the antenna to the resulting current flowing in the antenna I*, i.e.

$$Z_a = \frac{V(0)}{I_{\text{antenna}}} \tag{1.23}$$

where

- If $I_{\text{antenna}} = I_{\text{max}}$, then the impedance Z_a is referred to as the *loop current*;

- If $I_{\text{antenna}} = I(0)$, then the impedance Z_a is referred to as the *base current*;

Referring the impedance to the base current, it is written as

$$Z_a = \frac{V(0)}{I(0)} = R_a + jX_a \tag{1.24}$$

where X_a is the antenna reactance and $R_a = R_r + R_l$ is the antenna resistance, where R_r is the radiation resistance and R_l is the ohmic loss occurring in the antenna.

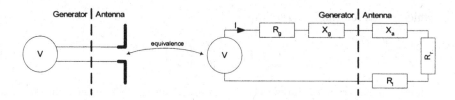

Fig. 1.4 Equivalence between physical radiating antenna and linear circuit.

1.4.6   Equivalent Circuit

The behaviour of a radiating antenna has therefore been reduced to that of an equivalent impedance. This is not a coincidence as Maxwell's equations treat both radiated EM-waves and guided EM-waves equally. The duality allows us to introduce an equivalent circuit as depicted in figure 1.4. It depicts a generator, here a voltage source, with internal impedance consisting of resistance R_g and reactance X_g being connected to the antenna with impedance consisting of resistance R_a and reactance X_a.

1.4.7   Antenna Matching

The equivalent circuit allows us to tune the impedance of the generator so as to maximise the radiated power delivered from the generator. This procedure is also often referred to as *impedance matching*. The average power delivered to the antenna is clearly given as

$$P = \frac{1}{2} I^2 R_a \tag{1.25}$$

where

$$I = \frac{V_g}{(R_a + R_g) + j \cdot (X_a + X_g)} \tag{1.26}$$

The power in (1.25) is maximised if conjugate matching is deployed, i.e. $R_g = R_a$ and $X_g = -X_a$, which yields a delivered power of

$$P = \frac{1}{8} \frac{V_g^2}{R_a} \tag{1.27}$$

1.4.8   Effective Length and Area

The effective length, l_e, *characterises the antenna's ability to transform the impinging electric field E into a voltage at the feeding point V(0), and vice versa,* and it is defined as

$$l_e = \frac{V(0)}{E} \tag{1.28}$$

which can be interpreted as the more voltage that is induced with less electric field strength, the bigger the effective length of the antenna.

The effective area, A_e, *characterises the antenna's ability to absorb the incident power density w and to deliver it to the load*, and it is defined as

$$A_e = \frac{P_{\text{load}}}{w}$$

(1.29)

which can be interpreted as the higher the delivered power with respect to the incident power density, the higher the effective area.

1.4.9 Radiation Intensity

The radiation intensity, U, is defined as *the power P per solid angle Ω*, and it is defined as

$$U = \frac{dP}{d\Omega}$$

(1.30)

This can be rewritten as

$$U = \frac{\text{Re}\left\{\mathbf{S}\right\} d\sigma}{d\Omega} = \frac{\text{Re}\left\{\mathbf{S}\right\} r^2 \, d\Omega}{d\Omega} = \text{Re}\left\{\mathbf{S}\right\} r^2$$

(1.31)

which, for the Hertzian dipole, yields

$$U = \frac{\eta I^2 \sin^2(\theta)}{8} \left(\frac{\delta l}{\lambda}\right)^2$$

(1.32)

and is clearly independent of the distance of observation.

1.4.10 Radiation Pattern

Since the radiation intensity U is independent of the distance of observations but only depends upon the antenna's inherent parameters, it can be taken to describe the radiation pattern of an antenna, an example of which is depicted in figure 1.5(a). If this radiation pattern is rolled out and depicted on a cartesian coordinate system, it will show a graph as depicted in figure 1.5(b). With reference to the example figure, the following is defined:

- *Radiation null:* Angle at which the radiated power is zero.

- *Main lobe:* The angular region between two radiation nulls which contains the angle with the strongest radiation intensity U_{max}.

- *Side lobes:* The angular regions between two radiation nulls which do not contain the angle with the strongest radiation power.

- *Half-power beamwidth (HPBW):* The angle spanned by the intensity region for which $U_{max}/2 \leq U \leq U_{max}$. The HPBW is associated with the ability of an antenna to direct a beam. The HPBW is often referred to as the 3 dB beamwidth for obvious reasons.

- *First null beamwidth (FNBW):* The angle spanned by the main lobe. The FNBW is associated with the ability of an antenna to reject interference.

- *Sidelobe level (SLL):* The power of the highest sidelobe relative to the peak of the main beam.

1.4.11 Bandwidth

The bandwidth B is defined as the *frequency band ranging from f_{lower} to f_{upper} within which the performance of the antenna, with respect to some characteristics, conforms to a specified standard, e.g. a drop by 3 dB.* Such a definition is fairly broad and includes characteristics such as radiation pattern, beamwidth, antenna gain, input impedance, and radiation efficiency. An antenna is said to be

- *narrowband* if $B_{narrow} = (f_{upper} - f_{lower})/f_{centre} < 5\%$;

- *wideband* if $f_{upper} : f_{lower} > 10 : 1$; and

- *frequency independent* if $f_{upper} : f_{lower} > 40 : 1$.

For example, an antenna is designed such that it radiates a total power of 0 dBm at a centre frequency of 1.8 GHz. It can be said that the specified standard is a power drop of 3 dB. Therefore, if the radiated power does not drop to -3 dBm within $f_{lower} = 1.755$ GHz and $f_{lower} = 1.845$ GHz but falls below -3 dBm out of that range, then the antenna is narrowband. If, however, the radiated power does not fall below -3 dBm well beyond 18 GHz, then the antenna is wideband.

1.4.12 Directive Gain, Directivity, Power Gain

An isotropic radiator is defined as *a radiator which radiates the same amount of power in all directions.* It is a purely hypothetical radiator used to aid the analysis of realisable antenna elements. For an isotropic radiator, the radiation intensity, U_0, is defined to be $U_0 = P/(4\pi)$.

This allows us to define the directive gain, g, as *the ratio of the radiation intensity U of the antenna to that of an isotropic radiator U_0 radiating the same amount of power,* which can be formulated as

$$g = \frac{U}{U_0} = 4\pi \frac{U}{P} \tag{1.33}$$

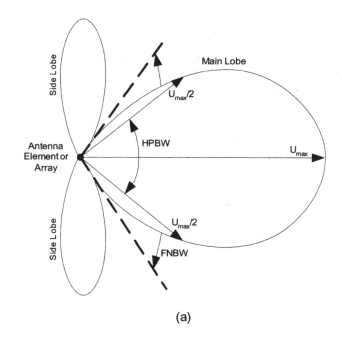

(a)

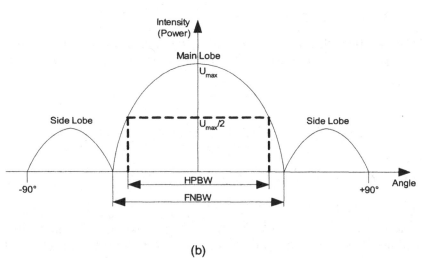

(b)

Fig. 1.5 Two different representations of a radiation pattern: (a) polar plot, (b) Cartesian plot.

For the Hertzian dipole,

$$g(\theta) = 1.5\sin^2(\theta) \tag{1.34}$$

which is clearly a function of direction, but not distance from the radiator.

From the above, the directivity D is defined as *the ratio of the maximum radiation intensity $U_{\max}$ of the antenna to that of an isotropic radiator U_0 radiating the same amount of power*, giving

$$D = \frac{U_{max}}{U_0} = 4\pi\frac{U_{\max}}{P} \tag{1.35}$$

and For the Hertzian dipole

$$D = 1.5 \tag{1.36}$$

which is simply a value and not a function of direction anymore.

This allows us finally to define the power gain G as *the ratio of the radiation intensity U of the antenna to that of an isotropic radiator U_0 radiating an amount of power equal to the power accepted by the antenna*, i.e.

$$G = \frac{U}{U_{0,input}} = 4\pi\frac{U}{P_{input}} \tag{1.37}$$

1.4.13   Radiation Efficiency

The radiation efficiency, e, is defined as *the ratio of the radiated power P to the total power P_{input} accepted by the antenna, where $P_{input} = P + P_{loss}$*, i.e.

$$e = \frac{P}{P_{input}} = \frac{P}{P + P_{loss}} \tag{1.38}$$

which can also be related to previous antenna parameters as $e = G/g$.

1.5   BASIC ANTENNA ELEMENTS

Any wire antenna can be viewed as the superimposition of an infinite amount of Hertzian dipoles which, theoretically speaking, allows one to calculate the resulting EM-field at any point in space by adding the field contributions of each Hertzian dipole. From (1.12) and (1.13) it has been shown that to calculate the field contributions, the current (and its direction) through the Hertzian dipole should be known.

For many antenna configurations, the current distribution can be calculated or estimated and the aforementioned theory applied. In this case, the antennas are referred to as *wire antennas*. In many other configurations, however, it is difficult to obtain an exact picture of the current distribution. In this case, it is easier to utilise Huygen's principle and deduce the radiated EM-field at any point in space from an estimate of the EM-field at a well-defined surface; such antennas are also referred to as *aperture antennas*.

Basic wire antennas include dipoles of finite length, loop and helix antennas. Basic aperture antennas include the horn and slot antenna, as well as parabolic dishes. Examples of common antenna elements are shown in figure 1.6. To obtain a feeling for the properties of these basic antenna elements, some of the most important are briefly considered in the following sections.

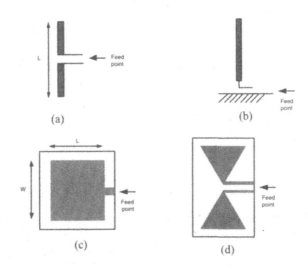

Fig. 1.6 Common types of antenna elements: (a) dipole, (b) mono-pole, (c) square patch, (d) bow tie.

1.5.1 Finite-Length Dipole

Dipoles of finite length L are of practical interest, an example of which is depicted in figure 1.6(a). If it is assumed that the dipole is fed with a sinusoidal voltage generator, then the resulting current in the finite-length dipole will also be (approximately) harmonic with a maximum amplitude of I_{max}. Such a feeding procedure is referred to as *balanced*, i.e. one feeding wire carries a current that is in anti-phase to the current in the other. That allows us to integrate over the field

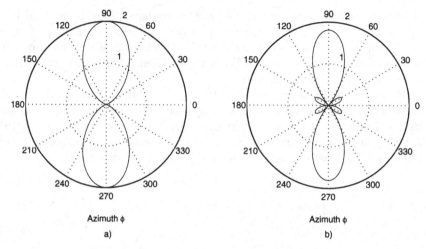

Fig. 1.7 Radiation pattern of a dipole of length: (a) $L = 1.0\lambda$ and (b) $L = 1.2\lambda$.

contributions (1.12) and (1.13) to arrive at [9]

$$E_\theta = \eta\, H_\phi \tag{1.39}$$

$$H_\phi = \frac{j I_{\max} e^{-jkr}}{2\pi r} \cdot P(\theta) \tag{1.40}$$

where $P(\theta)$ is the pattern factor given by

$$P(\theta) = \frac{\cos\left(\frac{1}{2}kL\cos(\theta)\right) - \cos\left(\frac{1}{2}kL\right)}{\sin(\theta)} \tag{1.41}$$

With (1.39) and (1.40), it is possible to calculate the antenna parameters introduced in section 1.4.

For example, the radiation resistance of a $\lambda/2$–dipole, i.e. $L = \lambda/2$, can be calculated to be $R_r = 73\Omega$ and the directivity is $D = 1.64$. Also, whilst radiating power uniformly over the azimuth plane, the finite-length dipole develops some interesting radiation patterns over the elevation plane. It can be shown that for $L < 1.1\lambda$ only one main lobe exists (figure 1.7(a), whereas for $L \geq 1.1\lambda$ the radiation pattern develops multilobes (figure 1.7(b). From this figure it is also observed that the power radiated along the main lobe decreases once multilobes develop, which is detrimental in most wireless applications. Note that the variation in radiation patterns observed between figures 1.7(a) and (b) also depict the change that occurs with frequency, since the radiation characteristics depend on the ratio between the physical length L and the wavelength λ.

1.5.2 Mono-pole

A mono-pole antenna is often considered as half of a dipole placed above a ground-plane, where the ground-plane acts as an electric mirror thus creating the other half of the dipole, as illustrated in figure 1.8. In contrast to the balanced feed for the dipole, a mono-pole is fed with a single-ended feeder where one wire carries the signal to the antenna and the other is connected to the ground-plane. This has the advantage of being directly connected to most receiver and transmitter modules that are often designed around a single-ended grounding system, and is fully compatible with co-axial cables and associated connectors.

Since half of the radiating plane is cut-off by the ground plane, the radiated power (and hence the radiation resistance) of a mono-pole is only half compared to the dipole with the same current; however, the directivity is doubled. Therefore, a $\lambda/4$ mono-pole has a radiation resistance of $R_r = 36.5\Omega$ and a directivity of $D = 3.28$. The latter, together with the compact spatial realisation and the simple feeding mechanism, is the main incentive to use mono-poles.

Fig. 1.8 Mono-pole above a ground-plane.

1.5.3 Printed Antennas

The antennas discussed so far have required a wire structure as a convenient way of realisation. Alternatively antennas can be produced using printed circuit techniques where one side of a copper clad dielectric is etched to the desired shape and the other is left as the ground-plane. Such antennas are often referred to as *printed antennas*. Dipoles can also be produced in this way, but this section focuses on patch antennas.

Patch antennas are practical and popular owing to their ease of manufacturing and the flexibility of design in terms of shape and topology. Many applications require antennas which are capable of conforming to the shape of the surface onto which they are mounted, and are required to have a planar profile for aesthetic or mechanical reasons. Patch antennas are a good solution for such applications. Figure 1.6(c) shows an example of a square patch antenna. Although the surface conductor can be practically any shape, rectangular and circular tend to be the most common.

The far-field radiation pattern is orientated orthogonal to the surface conductor, so in figure 1.6(c) it projects towards the user. As a rule of thumb, length L is approximately $\lambda_g/2$ and controls the operating frequency and width W is $0.9\lambda_g$ and controls the radiation resistance [10], where $\lambda_g = 1/\sqrt{\varepsilon_r}$ which is defined as the *normalised* wavelength in a media with relative permeability ε_r.

Note that contrary to common belief, the surface conductor does not form the radiating element as it does in a dipole. Instead, radiation occurs from along edges L and W, and which edge depends upon the electromagnetic mode of radiation the antenna is operating in. The radiation pattern of a square patch operating in the TE_{10} mode is $E_\phi \approx \cos(\theta)$.

It is evident from figure 1.6(c) that a single-ended feeder is required, as described in section 1.5.2 for the mono-pole antenna, and therefore has the characteristics associated with a single-ended feeder. Furthermore, although patch antennas have the advantages mentioned, they have a narrow operating bandwidth and low radiation efficiency compared to a dipole. Advanced structures can improve these parameters, an example of which is the stacked patch, which consists of several layers sandwiched together, where the size of each layer and the distance between the layers is carefully chosen.

1.5.4   Wideband Elements

With reference to the definition given in section 1.4.11, an antenna is said to be wideband if given characteristics, e.g. input impedance, do not change over a frequency band from f_{lower} to f_{upper} where $f_{upper} : f_{lower} > 10 : 1$. Simple antenna elements, such as the Hertzian dipole, finite-length dipole and mono-pole, are not capable of maintaining any characteristics over such wide bandwidths. More sophisticated antenna structures are hence required, an example of which are bow-tie antennas and horn antennas.

From previous analysis, it is shown that radiation characteristics depend on the ratio between the physical length L and the wavelength λ. From this fact, Rumsey observed that if an antenna design is only described by angles and is itself infinite in length, then it is inherently self-scaling and thus frequency-independent [9].

An example of a radiating structure obeying Rumsey's principle is depicted in figure 1.9(a). Since it is impossible to build the depicted radiating structure, which is infinite in size, several techniques have been suggested to make them effectively infinite. One of them is simply to truncate the infinite sheet, leading to a radiating

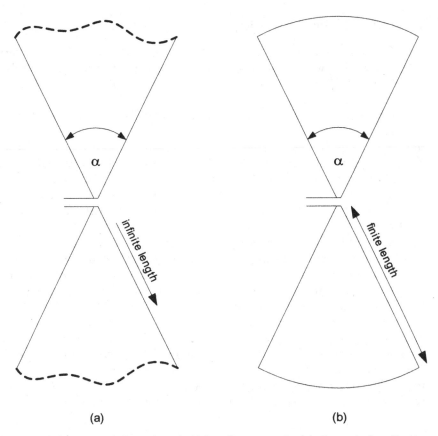

(a)   (b)

Fig. 1.9 A frequency-independent (wideband) antenna in (a) theoretical realisation and (b) practical realisation (bow-tie antenna).

structure as depicted in figure 1.9(b); such an antenna is also referred to as a *bow-tie antenna*. Since Rumsey's design properties are violated, the antenna will not be frequency-independent anymore; however, it exhibits parameter stability over a very wide frequency band, where exact numbers depend on the specific antenna realisation.

More sophisticated antenna elements are the *log-period toothed antenna*, *log-period trapezoidal wire antenna* and *log-period dipole array*. Due to their log-period self-scaling nature, they extend the effective length of the antenna structure. Finally, antennas which, by nature, are self-scaling, are fractal antennas.

1.5.5 Dual Polarised Elements

In section 1.3, the polarisation states of linear, circular and elliptical polarisation were introduced, which can be generated by means of two dipoles displaced by 90°. Such a radiation structure is referred to as a *dual polarised element* and is depicted in figure 1.10.

In addition to the polarisation states, it is possible to distinguish horizontal and vertical polarisation, which refers to the orientation of the polarisation with respect to some reference plane, such as ground or another surface. Therefore, if a linearly polarised wave has its E-field vector oscillating in a plane vertical to the ground, then such polarisation is referred to as *vertical polarisation*. If, on the other hand, the plane of the E-field vector oscillates in a plane parallel to the ground, then such polarisation is referred to as *horizontal polarisation*. Since the EM-field is linear, it is possible to decompose a field of any polarisation into vertically and horizontally polarised waves with respective phase differences.

Polarisation can be exploited in wireless systems to give operational advantages and therefore antenna elements are required that will efficiently transmit and receive electromagnetic waves with the required polarisation. Such antennas can be employed as sensors in adaptive antenna systems.

The dipole in figure 1.1 will generate or receive a vertically polarised electromagnetic wave with respect to the ground. By rotating the antenna through 90°, the antenna is orientated for horizontal polarisation. Therefore the simplest way to simultaneously operate in vertical and horizontally polarised modes is to employ two orthogonal dipoles as shown in figure 1.10. Notice that this configuration has two feedpoints, one for each polarisation.

Fig. 1.10 Dual polarised dipole.

Patch antennas can conveniently be configured for dual polarised operation by simply employing an additional feed point for the second polarisation at the appropriate location.

Circular polarisation operation is obtained in the same manner as linear polarisation except that appropriate phasing of the two waves is required between the two feed points, i.e. 90°.

1.5.6 Sonar Sensors

Sonar systems are a water-borne equivalent of radar systems. They operate using transverse waves. Due to this difference, they require different sonar sensors to receive and transmit the signals. Note, also, that the operative wavelength in the acoustic spectrum becomes comparable to a radar operating in the microwave EM spectrum.

Sonar systems are often classified as either *active* or *passive*. Active systems employ a transmitter and receiver and send out a signal which is subsequently detected by the receiver and processed to give information about the environment, e.g. other vessels or fish. Passive sonar only consists of a receiver which monitors the surrounding signal sources.

The receiving microphone is termed a *hydrophone* and the transmitting loudspeaker is termed a *projector*. In the case of a sensor having the dual purpose of transmitting and receiving, the sensor is referred to as a *transducer*.

The output power from the transducer is termed *source power level* (SPL). Like antenna elements, transducers can be designed to have directional properties and a 3 dB beamwidth, θ. The beamwidth can only be made as small as the diffraction limit of the transducer, which is a function of the diameter, D, of the transducer's aperture. The beamwidth is given by:

$$\theta = \frac{2\lambda}{D} \tag{1.42}$$

For example, assuming a transducer operating in sea water with $D = 10$ cm and $\lambda = 15$ cm, then a 3 dB-beamwidth of $\theta = 3°$ is obtained.

1.6 ANTENNA ARRAYS

From section 1.4.9, it can be shown that the HPBW of a Hertzian dipole is $90°$. In most wireless terrestrial and space applications, a narrower HPBW is desired because it is desirable to direct power in one direction and no power into other directions. For example, if a Hertzian dipole is used for space communication, only a fraction of the total power would reach the satellite and most power would simply be lost (space applications require HPBWs of well below $5°$).

With a finite-length dipole, it would be possible to decrease the HPBW down to $50°$ by increasing the length to $L \approx 1.1\lambda$; a further increase in length causes the HPBW to decrease further; however, unfortunately multilobes are generated hence further decreasing the useful power radiated into the specified direction.

It turns out that with the aid of antenna arrays, it is possible to construct radiation patterns of arbitrary beamwidth and orientation, both of which can be controlled electronically. An *antenna array* is by definition a radiating configuration consisting of more than one antenna element. The definition does not specify which antenna elements are used to form the array, nor how the spatial arrangement ought to be. This allows us to build antenna arrays consisting of different elements, feeding arrangements and spatial placement, hence resulting in radiating structures of different properties.

Example antenna arrays consisting of several patch antennas are depicted in figure 1.11, and briefly dealt with below.

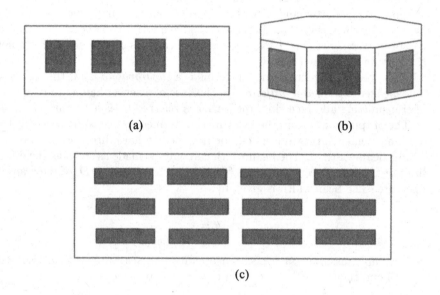

(a)

(b)

(c)

Fig. 1.11 Common types of antenna arrays: (a) linear array, (b) circular array, (c) planar array.

1.6.1   Linear Array

The most common and most analysed structure is the *linear antenna array*, which consists of antenna elements separated on a straight line by a given distance. Although each single antenna element can have a large HPBW, the amplitude and phase of the feeding current to each element can be adjusted in a controlled manner such that power is transmitted (received) to (from) a given spatial direction.

If adjacent elements are equally spaced then the array is referred to as a *uniform linear array* (ULA). If, in addition, the phase α_n of the feeding current to the nth antenna element is increased by $\alpha_n = n\alpha$, where α is a constant, then the array is a *progressive phaseshift array*. Finally, if the feeding amplitudes are constant, i.e. $I_n = I$, then this is referred to as a *uniform array*.

The uniform array, depicted in figure 1.11(a), with an inter-element spacing of $\lambda/2$ is the most commonly deployed array as it allows simple feeding, beamsteering and analysis. For example, the power radiated in azimuth at an elevation of $\theta = \pi/2$ for such an array consisting of $\lambda/2$-dipoles can be calculated as

$$P(\theta = \pi/2, \phi) \propto \left| \frac{1}{N} \frac{\sin(\frac{1}{2}N(\pi \cos(\phi) + \alpha))}{\sin(\frac{1}{2}(\pi \cos(\phi) + \alpha))} \right|^2 \qquad (1.43)$$

where α is the progressive phase shift, N is the total number of antenna elements, and it has been assumed that $I_n = 1/N$.

Figure 1.12 depicts the dependency of the radiation pattern on the number of antenna elements, where $\alpha = 0$. It can be observed that increasing the number of elements N, also decreases the HPBW where the width is inversely proportional to the number of elements. It is also observed that, independent of the number of elements, the ratio between the powers of the main lobe and the first side lobe is approximately 13.5 dB. Therefore, if such a level does not suit the application, antenna arrays different from uniform arrays have to be deployed.

Figure 1.13 illustrates the dependency of the beam direction on the progressive feeding phase α, where $N = 10$. Increasing α increases the angle between the broadside direction and the main beam, where the dependency is generally not linear.

In summary, by means of a simply employable uniform antenna array, the HPBW can be reduced by increasing the number of antenna elements and steering the main beam into an arbitrary direction by adjusting the progressive phase shift.

1.6.2 Circular Array

If the elements are arranged in a circular manner as depicted in figure 1.11(b), then the array is referred to as an *uniform circular array* (UCA). With the same number of elements and the same spacing between them, the circular array produces beams of a wider width than the corresponding linear array. However, it outperforms the linear array in terms of diversity reception.

If maximal ratio combining is used, then it can be shown that the UCA outperforms the ULA on average for small and moderate angle spread for similar aperture sizes. However, the ULA outperforms the UCA for near-broadside angles-of-arrival with medium angular spreads. It can also be shown that the central angle-of-arrival has a significant impact on the performance of the ULA, whereas the UCA is less susceptible to it due to its symmetrical configuration.

1.6.3 Planar Array

Both linear and circular arrangements allow one to steer the beam into any direction in the azimuth plane; however, the elevation radiation pattern is dictated

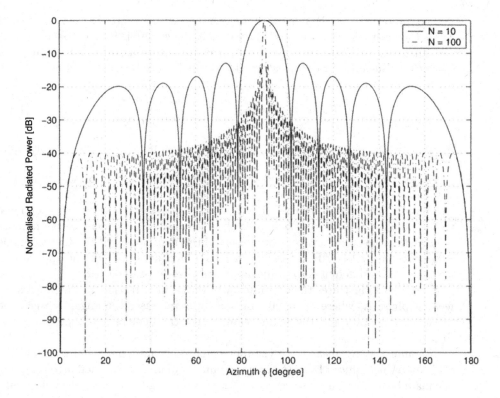

Fig. 1.12 Dependency of the radiation pattern on the number of elements N.

by the radiation pattern of the antenna elements. In contrast, the *planar antenna array* allows one also to steer the beam in elevation, thereby producing so-called *pencil beams*. An example realisation of a planar array is depicted in figure 1.11(c). For the planar antenna array, the same observations as for the linear array hold.

1.6.4   Conformal Arrays

The array types discussed so far have all been based upon a regular, symmetrical design. This is suitable for applications where such mounting is possible; however, this is not the case for many scenarios where the surface is irregular or the space is confined. For these situations a conformal array is required which, as the name suggests, conforms to the surrounding topology. The arrays could be elliptical or follow a more complex topology. The challenge for the design of these antennas is to make sure that the main lobe beamwidth and side lobe levels fall within the required specification. Specialised antenna elements are also required since

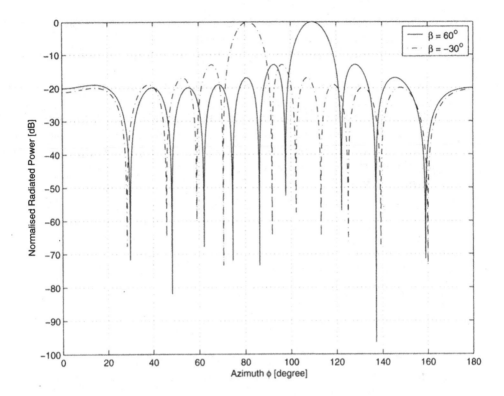

Fig. 1.13 Dependency of the beam direction on the progressive feeding phase α.

they will be mounted on an irregular surface and must therefore follow the surface contours. This particularly applies to patch antennas which usually require a flat surface.

1.7 SPATIAL FILTERING

From the previous section, it has been shown that appropriate feeding allows antenna arrays to steer their beam and nulls towards certain directions, which is often referred to as *spatial filtering*. Spatial filtering is of particular importance in mobile communication systems since their performance tends to be interference limited.

The very first mobile communication systems had base stations with an omnidirectional antenna element, i.e. the transmit power was equally spread over the entire cell and hence serving everybody equally. However, since many such communication cells are placed beside each other, they interfere with each other. This is because a mobile terminal at a cell fringe receives only a weak signal from its own base station, whereas the signals from the interfering base stations grow stronger.

Modern wireless communication systems deploy antenna arrays, as depicted in figure 1.14. Here, a base station communicates with several active users by directing the beam towards them and it nulls users which cause interference. This has two beneficial effects: first, the target users receive more power compared to the omnidirectional case (or, alternatively, transmit power can be saved); and, second, the interference to adjacent cells is decreased because only very selected directions are targeted.

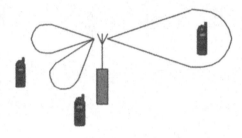

Fig. 1.14 Principle of spatial filtering applied to a mobile communication system.

The above-described spatial filtering can be realised by means of the following mechanisms:

- **Sectorisation:** The simplest way is to deploy antenna elements which inherently have a sectorised radiation pattern. For instance, in current second-generation mobile phone systems, three base station antennas with 120° sectors are deployed to cover the entire cell.

- **Switched beam:** A beam is generated by switching between separate directive antennas or predefined beams of an antenna array. Traditionally, algorithms are in place which guarantee that the beam with the strongest signal is chosen.

- **Phased antenna array:** By adjusting the feeding phase of the currents, a moveable beam can be generated (see section 1.6.1 and figure 1.13). The feeding phase is adjusted such that the signal level is maximised.

- **Adaptive antenna array:** As with the phased array, a main lobe is generated in the direction of the strongest signal component. Additionally, side lobes are generated in the direction of multipath components and also interferers are nulled. Here, not only the signal power is maximised but also the interference power is minimised, which clearly requires algorithms of higher complexity to be deployed when compared to the switched beam and phased arrays. Since adaptive antenna arrays are currently of great importance and also the main topic of this book, their functioning is briefly introduced below.

1.8 ADAPTIVE ANTENNA ARRAYS

Why adaptive? Clearly, adaptability is required to change its characteristics to satisfy new requirements, which may be due to

- changing propagation environment, such as moving vehicles;

- changing filtering requirements such as new targets or users requiring processing.

This is particularly the case for spatial filtering which can dynamically update the main beam width and direction, side-lobe levels and direction of nulls as the filtering requirement and/or operating environment changes. This is achieved by employing an antenna array where the feeding currents can be updated as required. Since the feeding current is characterised by amplitude and phase, it can be represented by a complex number. For this reason, from now on explicit reference to the feeding current is omitted and instead we refer to the *complex array weights*. Such a system with adaptive array weights is shown in figure 1.15. Clearly, the aim of any analysis related to adaptive antenna arrays is to find the optimum array weights so as to satisfy a given performance criterion.

1.9 MUTUAL COUPLING & CORRELATION

Any form of beamforming relies on the departure of at least two phase-shifted and coherent waves. If either the phase-shift or the coherence is violated, then no beamforming can take place; this may happen as follows.

If two antenna elements are placed very closely together, then the radiated EM field of one antenna couples into the other antenna and vice versa and therefore erasing any possible phase-shift between two waves departing from these elements; this effect is known as *mutual coupling*. The antenna array designer has hence to make sure that the mutual coupling between any of the elements of an antenna array is minimised. This is usually accomplished for inter-element spacing larger than $\lambda/2$. If, on the other hand, the spacing is too large, then the departing waves do not appear coherent anymore, hence, also preventing an EM wave to form a directed beam.

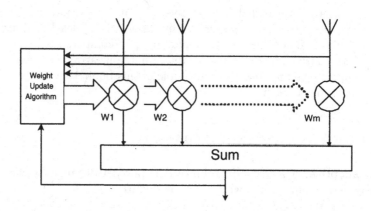

Fig. 1.15   Concept of an adaptive antenna array.

From a purely antenna array point of view, a suitable inter-element spacing for beamforming is thus in the order of $\lambda/2$. Sometimes, however, the antenna array is not utilised as a beamformer, but as a diversity array. This requires the arriving and departing signals to be as de-correlated as possible. It can be shown that if the antenna array is uniformly surrounded by clutter, then the spatial correlation function of the channel requires a minimal inter-antenna element spacing of $\lambda/2$ for the signals to appear de-correlated. Such ideal clutter arrangement, however, is rarely found in real-world applications where the correlation between antenna elements increases as the angular spread of the impinging waves decreases. For instance, a base station mounted on the rooftop with little clutter around requires spacings of up to 10λ to achieve de-correlation between the elements.

From this short overview, it is clear that the antenna array designer faces many possible realisations where the optimum choice will depend on the application at hand.

1.10   CHAPTER SUMMARY

This chapter has introduced the most basic concepts required to understand the remainder of this book. The functioning of an antenna element has been explained by means of Maxwell's equations. This facilitated the analytical description of the radiation properties of the smallest possible element, the infinitesimal small Hertzian dipole. Some antenna parameters have then been introduced and quantified for the Hertzian dipole, but which are also applicable to any other form of radiating element.

Typically occurring antenna elements have then been described, which can be viewed as a superposition of an infinite number of Hertzian dipoles. These elements include the finite length half-wavelength dipole, the mono-pole and other radiating structures.

Placed together, these antenna elements form antenna arrays, and distinction has been made between linear, circular and planar antenna arrays. Their advantages and disadvantages have also been discussed. Finally, the antenna arrays were applied to achieve spatial filtering by means of adapting the feeding currents.

1.11 PROBLEMS

Problem 1. Why is the Hertzian dipole the basis of any radiating structure?

Problem 2. Prove equations (1.12) and (1.13) from equations (1.9)–(1.11).

Problem 3. Show that equation (1.17) holds by utilising equations (1.12) and (1.13).

Problem 4. Demonstrate that the complex conjugate matching indeed maximises the radiated power, and show that equation (1.27) holds.

Problem 5. What is the difference between the bandwidth and beamwidth of an antenna?

Problem 6. With the help of equation (1.43), show that the ratio between the mainlobe and first sidelobe is approximately 13.5 dB and, for N being large, independent of the number of antenna elements N forming the array.

Problem 7. Explain the advantage of employing an adaptive antenna in a mobile communications system.

Problem 8. Explain why 'adaptivity' is fundamental to successful tracking of a missile using an antenna array.

Problem 9. Explain how an array may improve the performance of a sonar system.

2

Narrowband Array Systems

2.1 INTRODUCTION

In the previous chapter sensors and antenna elements were described and shown how they can be assembled into an array of various topologies. Consequently, antenna arrays were introduced where a beam can be steered by multiplying the received signal at each element by a complex weight. The weights are chosen to steer the beam in a certain direction. The weighted signals are finally summed to form a single signal available for the application to process further. The weights are chosen such that the combined signal is enhanced in the presence of noise and interference.

This chapter builds on this concept by initially introducing essential terminology, then demonstrating the function of the complex weights and how the phase and amplitude weights impact on the control of the array pattern. For example, the main beam can be steered toward a target for signal enhancement, or nulls can be steered toward interferers in order to attenuate them. Grating lobes are also investigated and it is shown how these can produce sporadic effects when the array is used as a spatial filter. It is then shown how the application of window functions can be used to control the sidelobes of the radiation pattern by controlling the trade-off between the main-beam width and the sidelobe level. Thus, very low sidelobes can be achieved in order to give good rejection to signals arriving at these angles at the expense of having a wider main-beam.

The focus of this chapter is narrowband beamformers. The term *narrowband* has many definitions. For clarity, it is defined here as an array operating with

Adaptive Array Systems   B. Allen and M. Ghavami
© 2005 John Wiley & Sons, Ltd   ISBN 0-470-86189-4

signals having a fractional bandwidth (FB) of less than 1% i.e., FB < 1%, where FB is defined by the following equation

$$FB = \frac{f_h - f_l}{(f_h + f_l)/2} \times 100\% \tag{2.1}$$

where f_h and f_l are the highest and lowest components of the signal respectively. As the bandwidth of the signal is increased the phase weighting computed to steer the beam at a particular angle is no longer correct, therefore a more complex weighting arrangement is required. Wideband array processing techniques i.e., array processing for signals with FB > 1% are described in chapter 3.

2.2   ADAPTIVE ANTENNA TERMINOLOGY

In order that beamforming techniques can be analysed, fundamental terminology must first be explained. This section introduces basic terms such as *steering vector* and *weight vector*.

Consider an array of L omni-directional elements operating in free space where a distant source is radiating sinusoids at frequency f_o, as shown in figure 2.1 for $L = 2$, i.e., a two-element array. The signals arriving at each of the sensors will be both time and phase delayed with respect to each other due to the increased distance of the sensors from a reference sensor. This is evident from the figure where the wave front will arrive at element 2 before element 1. This means that the phase of the signal induced in element 2 will lead that in element 1. The phase difference between the two signals depends upon the distance between the elements, d, and the angle of the transmitter with respect to the array boresight, shown as θ in the figure. The additional distance the wave has travelled to arrive at element 1, Δd, is given by

$$\Delta d = d \cdot \sin \theta \tag{2.2}$$

If a planar array is considered with elements in the x-y plane, the signal arriving at each of the sensors will therefore arrive at angles θ and ϕ, and in this more general case, Δd becomes

$$\Delta d = d \cdot \sin \phi \sin \theta \tag{2.3}$$

and the phase difference becomes

$$\Delta \Phi = 2\pi \frac{d}{\lambda} \sin \phi \sin \theta \tag{2.4}$$

where λ is the wavelength. This signal model assumes that the wave arrives at each element with equal amplitude.

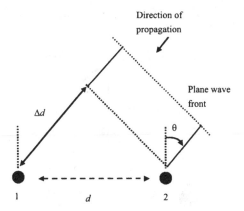

Fig. 2.1 Array signal model for a two-element array. d = distance between elements, Θ = angle of arrival of wave from boresight, Δd = additional distance of wave.

After the signals have arrived at the sensors, they are weighted and subsequently summed. Consider a simple array consisting of two elements with unity amplitude and zero phase weighting, the received signal after summation is given by the equation below, where element 1 is the phase reference point.

$$E(\theta, \phi) = 1 + e^{j2\pi \frac{d}{\lambda} \sin \phi \sin \theta} \tag{2.5}$$

Plotting $10 \log_{10}(|E(\theta, \phi)|)$ as a function of θ and ϕ yields the far-field radiation pattern of the array. Note that the two terms in the above equation refer to the received signals from the transmitter at each of the elements with respect to the first element. The terms can be expressed as a vector, referred to as the *received signal vector*, thus the received signal vector for the above example is

$$\mathbf{r} = [\ 1 \quad e^{j2\pi \frac{d}{\lambda} \sin \phi \sin \theta} \ ] \tag{2.6}$$

For N equally spaced elements, equation (2.5) is generalised as:

$$E(\theta, \phi) = \sum_{n=1}^{N} e^{j2\pi \frac{d}{\lambda} n \sin \phi \sin \theta} \tag{2.7}$$

and can be easily extended to accommodate unequally spaced elements by including the distance of each element with respect to the first.

If several transmitters are detected by the sensors, the received signals can be expressed in the same way for each of the transmitters, thus each of the signals have an associated received signal vector, i.e., if there are K transmitters, K received signal vectors can be determined. This frequently occurs in practice, such

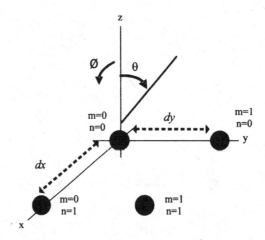

Fig. 2.2   Planar array signal model for $M = 2$, $N = 2$ array. $d_x = $ distance between x-axis elements, $d_y = $ distance between y-axis elements, $\theta = $ azimuth angle of arrival of wave from boresight, $\phi = $ elevation angle of arrival of wave from boresight.

as a radar tracking multiple targets, or in a mobile communications system where multiple users are active. The received signal vector of the kth signal is frequently referred to as the *steering vector*, $\mathbf{S_k}$.

The above analysis is generalised for a planar array consisting of $M \times N$ uniformly spaced elements and where the radiation pattern is evaluated in azimuth (θ) and elevation (ϕ) as shown in figure 2.2. The resulting radiation pattern is given by:

$$E(\theta, \phi) = \sum_{m=1}^{M} \sum_{n=1}^{N} e^{j \frac{2\pi}{\lambda}(md_x \cos\theta + nd_y \sin\theta \sin\phi)} \tag{2.8}$$

where d_x and d_y are the element spacing in the x and y dimensions respectively. Note that equating equation (2.8) to zero and solving for θ and/or ϕ gives the location of the nulls. For a two-element array, nulls occur in the pattern when

$$\cos\left(\pi \frac{d}{\lambda} \sin\theta \sin\phi\right) = 0 \tag{2.9}$$

$$\pi \frac{d}{\lambda} \sin\theta \sin\phi = \frac{2n + 1}{2}\pi \tag{2.10}$$

where $n = 0, 1, 2, \dots$ Assuming $\theta = 90°$,

$$\phi_{nulls} = \sin^{-1}\left(\left[n + \frac{1}{2}\right]\frac{\lambda}{d}\right) \tag{2.11}$$

Furthermore, the impact of applying weights to each of the received signals can also be analysed. The above analysis assumed unity amplitude and zero phase weights applied to each element, thus for a two-element array, the weights can be expressed as the vector $\mathbf{W} = [\ 1 \cdot e^{j0} \quad 1 \cdot e^{j0}\ ]$, which is referred to as the *weight vector*. The values of the weights are computed to give a desired response, i.e., to steer the main beam and nulls in the desired directions and set the sidelobes to the desired level and set the main beam width. The computation of $\mathbf{W}$ is discussed in the subsequent subsections. $\mathbf{W}$ is a $1 \times N$ vector for a linear array and a $M \times N$ matrix for a planar array. It is incorporated into the analysis by multiplying the array response by $\mathbf{W}$ as shown below.

$$E(\theta, \phi) = \sum_{m=1}^{M} \sum_{n=1}^{N} W_{mn} e^{j\frac{2\pi}{\lambda}(md_x \cos\theta + nd_y \sin\theta \sin\phi)} \tag{2.12}$$

2.3 BEAM STEERING

The primary function of an adaptive array is to control the radiation pattern so that the main beam points in a desired direction and to control the sidelobe levels and directions of the nulls. This is achieved by setting the complex weights associated with each element to values that cause the array to respond desirably.

The complex weights consist of real and imaginary components, or alternatively, amplitude and phase components, i.e., $A_n = \alpha_n e^{j\beta_n}$ where A_n is the complex weight of the nth element, α_n is the amplitude weight of the nth element and β_n is the phase weight of the nth element. The phase components control the angles of the main beam and nulls, and the amplitude components control the sidelobe level and main beam width (although when the main beam is steered towards end-fire by phase weighting, the main beam-width will also increase). These are considered in turn in the following sections.

2.3.1 Phase Weights

The main beam can be steered by applying a phase taper across the elements of the array. For example, by applying phases of $0°$ and $70°$ to each element of a two-element array with $\lambda/2$ spacing between the elements, the beam is steered to $35°$, as shown in figure 2.3. The figure also shows the radiation pattern when zero phase weights are applied, i.e., the main beam is steered towards boresight and a gain of 3 dB is shown. The following section describes a method for computing the required weights to steer the main beam to a desired angle.

Phase weights for narrowband arrays, which is the focus of this chapter, are applied by a phase shifter, However, where wideband beamforming is considered, which is the focus of chapter 4, it is important to distinguish between phase-shifter arrays and delay arrays. The first generates a phase shift for a certain frequency of

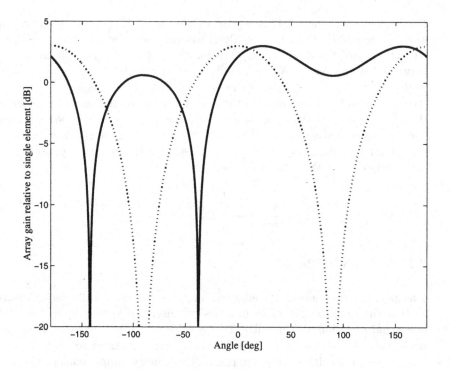

Fig. 2.3 Radiation pattern with phase weighting for a two-element array with $\frac{\lambda}{2}$ spacing. Dotted line with no phase tapper, solid line with beam steered to 35°.

operation and can be used for narrowband arrays operating over a very small range of frequencies. On the other hand delay arrays generate a pure time delay for each element depending on the desired angle and produce wideband characteristics. This is because in phase arrays, as the frequency varies, the steering angle of the array also changes, but in delay (wideband) arrays, the steering angle never changes with frequency. Of course, in the delay array sidelobe variations can be observed with frequency. Pure delays can be implemented by digital filters, analogue devices or optical circuits.

2.3.2   Main Beam Steering

A simple beamformer steers the main beam in a particular direction (θ, ϕ). The weight vector, $\mathbf{W}$, for steering the main beam is given by

$$\mathbf{W} = \frac{1}{L}\mathbf{S_o} \qquad (2.13)$$

where L is the number of elements and $\mathbf{S_o}$ is the steering vector. Due to the normalisation factor $\frac{1}{L}$, this beamformer yields a unity response in the look-direction (not considered in equation (2.12)). In other words, the output power is the same as that of a single element system. In an environment consisting only of noise, i.e., no interfering signals, this beamformer provides maximum SNR. This can be determined as follows. The autocorrelation function of the noise, $\mathbf{R}_N$, is given by:

$$\mathbf{R}_N = \sigma_N^2 \mathbf{I} \tag{2.14}$$

where σ_N^2 is the noise variance and $\mathbf{I}$ is the identity matrix. Thus the noise power at the output of the beamformer is

$$P_N = \mathbf{W}^H \mathbf{R}_N \mathbf{W} = \frac{\sigma_N^2}{L} \tag{2.15}$$

where $.^H$ is the Hermitian transpose. Thus the noise power is reduced by a factor of L.

Although this beamformer yields a maximum SNR in an environment consisting only of noise, this will not produce a maximum SNR in the presence of directional interference. Such scenarios are very common in radar and sonar deployments where an intentional jammer may be targeting the area, as well as mobile communications where other network users will create unintentional interference. The following section introduces a beamformer for scenarios where the number of directional interfering signals, $N \leq L - 1$.

Note that the above technique for computing the weights to steer the main beam has considered a two-dimensional array, should the beam of a planar array require steering in both azimuth and elevation, the technique can be applied for both planes.

Example 2.1

A receiver equipped with a four-element linear array is required to enhance the wanted signal arriving from a transmitter located at $20°$ from the array boresight. Compute the required weight vector for the array to perform this function given that the elements are spaced apart by $d = \lambda/2$.

The weight vector, $\mathbf{W}$, is computed using equation (2.13). Thus, $\mathbf{S_o}$ is computed to be

$$\mathbf{S_o} = [\ 1 \quad e^{j1.0745} \quad e^{j2.149} \quad e^{j3.223}\ ] \tag{2.16}$$

and from equation (2.13),

$$\mathbf{W} = \frac{1}{4}[\ 1 \quad e^{j1.0745} \quad e^{j2.149} \quad e^{j3.223}\ ] \tag{2.17}$$

2.3.3 Null Steering

As well as directing the main beam, the nulls in the array pattern can also be directed. This is useful when it is necessary to attenuate unwanted signals arriving at angles other than that of the main beam. This subsequently increases the signal-to-interference ratio at the output of the beamformer. Figure 2.3 illustrates how null steering would improve the performance of this system. With the main beam steered at $35°$, the radiation pattern shows that around 3 dB of rejection is achieved for signals arriving in the range of $50°$ to $100°$. Considering the additional cost and complexity of implementing this system, 3 dB rejection is not a significant improvement. However, by steering one of the nulls (currently at $-40°$ and $-140°$) to $90°$, 20 dB of rejection can be achieved for signals arriving within this angular range. This is a significant improvement for a small amount of computation.

Assume an array pattern with unity response in the desired direction and nulls in the directions of the interfering signals, the weights of the beamformer are required to satisfy these constraints [11, 12].

Let S_0 be the main beam steering vector and $S_1, \ldots, S_k$ are k steering vectors for the k nulls. The desired weight vector is the solution to the following simultaneous equations

$$\mathbf{W}^H \mathbf{S}_0 = 1 \qquad (2.18)$$

$$\mathbf{W}^H \mathbf{S}_k = 0 \qquad (2.19)$$

Let the columns of matrix $\mathbf{A}$ contain the $k + 1$ steering vectors and

$$\mathbf{c} = [\, 1 \quad 0 \quad \ldots \quad 0 \,]^T \qquad (2.20)$$

For $k = L - 1$, where L is the number of elements and $\mathbf{A}$ is a square matrix. The above equations can be written as

$$\mathbf{W}^H = \mathbf{A}\mathbf{C}_1^T \qquad (2.21)$$

Thus, the weight vector, $\mathbf{W}^H$, can be found by the equation below.

$$\mathbf{W}^H = \mathbf{C}_1^T \mathbf{A}^{-1} \qquad (2.22)$$

For $\mathbf{A}$ to exist, it is required that all steering vectors are linearly independent. Should this not be the case, the pseudo-inverse can be found in its place.

When the number of required nulls is less than $L - 1$, $\mathbf{A}$ is not a square matrix. Under such conditions suitable weights may be given by

$$\mathbf{W}^H = \mathbf{C}_1^T \mathbf{A}^H (\mathbf{A}\mathbf{A}^H)^{-1} \qquad (2.23)$$

However this solution does not minimise the uncorrelated noise at the array output.

Example 2.2

Compute the weight vector required to steer the main beam of a two-element array towards the required signal at $45°$ and to minimise the interfering signal that appears at $-10°$. Assume the element spacing $d = \frac{\lambda}{2}$. Such a scenario is typical of a TV receiver situated on a tall building and is therefore able to receive signals from both a local transmitter and a distant co-channel transmitter. A null is then steered towards the co-channel transmitter to provide attenuation of this unwanted signal.

The weight vector $\mathbf{W}$ is computed using equation (2.22). The steering vectors $\mathbf{S}_0$ and $\mathbf{S}_1$ are first required and computed with reference to equation (2.7) as follows.

$$\mathbf{S}_0 = [\ 1 \quad e^{-j2.221}\ ] \tag{2.24}$$

$$\mathbf{S}_1 = [\ 1 \quad e^{j0.5455}\ ] \tag{2.25}$$

Thus,

$$\mathbf{A} = [\ \mathbf{S}_0 \quad \mathbf{S}_1\ ] \tag{2.26}$$

and therefore,

$$\mathbf{A} = \begin{bmatrix} 1 & 1 \\ e^{-j2.221} & e^{j0.5455} \end{bmatrix}, \tag{2.27}$$

and $\mathbf{C} = [\ 1 \quad 0\ ]$
and after computing $\mathbf{A}^{-1}$, it is trivial to compute the weight vector to be

$$\mathbf{W} = [\ 0.5 - j0.095 \quad -0.3782 + j0.3405\ ] \tag{2.28}$$

The resulting radiation pattern with the main beam steered towards $45°$ and a null at $-10°$ is depicted in figure 2.4. The null is shown to provide an attenuation in excess of 20 dB. In this example, the angle between the null and the main beam is $55°$. This is comparable to the main beam width of a two-element array. If the null was required to be closer to the main beam it would interfere with the main beam shape therefore compromising the reception of the wanted signal. Cases such as this can be circumvented by increasing the number of antenna elements which would narrow the main beam therefore enabling a null to be formed closer to it. Increasing the number of elements would also:

- reduce the sidelobe level (up to a maximum of 13 dB);

- increase the main beam gain;

- increase the available degrees of freedom, and therefore the number of nulls; and

- also increase the cost.

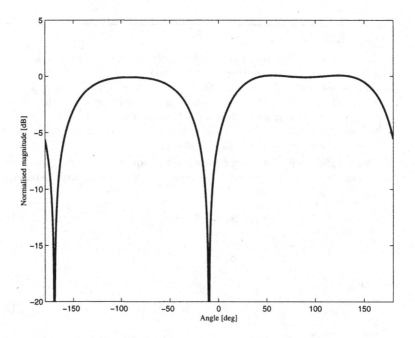

Fig. 2.4   Joint beam and null steering: Main beam steered to 45°, null steered to −10°.

If the number of elements is increased and only one null is required to be steered, **A** is no longer a square matrix and the array weights would then be computed by means of equation (2.23).

One of the drawbacks of null steering is the alignment of nulls in the desired directions due to their narrow widths. Exact alignment requires an accuracy of <2°, i.e., from the figure, the null width is 2° for a null depth of 10 dB. Thus, a small misalignment of the null of 0.5° will result in a 5 dB reduction of the attenuation of the co-channel signal. This can be overcome by null broadening which is simply achieved by steering multiple nulls towards the same target, but with a small angular offset between them [13]. This technique is particularly useful in mobile communications where the signal propagation conditions often cause the signals to disperse, as described in chapter 5. Furthermore, a deep null can only be achieved by a well-designed system and a practical system will compromise this due to system imperfections, although system calibration techniques can reduce the impact of imperfection to a certain extent. This is also discussed further in chapter 5.

The null steering technique described here jointly steers the main beam and nulls to the desired angles. Modifying the vector **C** enables the existence of nulls and beams (or signal minima and maxima) to be specified according to the prevailing requirements.

2.4   GRATING LOBES

So far the analysis of the behaviour of arrays has considered elements with a spacing of $d = \frac{\lambda}{2}$. What happens when other element spacings are used? Such a scenario occurs when the array is used over a wide frequency range and may also occur when certain physical design constraints are imposed on the array, such as element size. This is explored in the following example.

Assuming a two-element array with unity amplitude and zero-phase weights, the radiation pattern is obtained from equation (2.7). Plotting this expression for $d = 0.5\lambda, 0.6\lambda$ and 2λ gives the radiation patterns shown in figure 2.5. It can be seen that, as the element spacing becomes wider, the main lobe narrows and, when $d > 0.5\lambda$, additional lobes appear with an increasing amount of energy appearing in them as d is further increased. These additional lobes are referred to as *grating lobes*. The time domain analogy of grating lobes is aliasing. Aliasing occurs when the Nyquist sampling theorem is not obeyed. For the $\theta = 90°$ plane, the positions of these lobes are given by:

$$\cos\left(\pi\frac{d}{\lambda}\sin\phi_{\max}\right) \tag{2.29}$$

where

$$\pi\frac{d}{\lambda}\sin\phi_{\max} = n\pi \qquad n = 0, 1, 2\ldots \tag{2.30}$$

$$\phi_{\max} = \sin^{-1}\left(n\frac{\lambda}{d}\right) \tag{2.31}$$

Example 2.3

The following example illustrates the impact of grating lobes in a 3G (UTRA) mobile communications network [14].

A radiation pattern generated from a uniform linear antenna array will contain complete grating lobes if the following equation is satisfied for values other than $n = 0$ [15].

$$\pi\frac{d}{\lambda}[\sin\theta_g - \sin\theta_0] = \pm n\pi \qquad n = 0, 1, 2\ldots \tag{2.32}$$

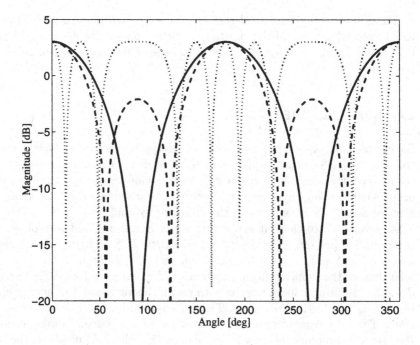

Fig. 2.5 Examples of grating lobes for $d = 0.5\lambda$, 0.6λ, 2λ. Solid line $d = 0.5\lambda$, dashed line $d = 0.6\lambda$, dotted line $d = 2\lambda$.

where:
θ_g = angular position of the grating lobe in azimuth plane
θ_o = angular position of the main lobe in azimuth plane
d = inter-element distance
λ = carrier wavelength

Assume the angular coverage of the cell is $120°$, the maximum angle the main beam can be steered towards is $60°$ i.e. $\theta_{0(\mathrm{max})} = \pm 60°$. For $n = 1$,

$$\sin \theta_g = 0.8667 \pm \frac{\lambda}{d} \qquad (2.33)$$

and in real space, $|\sin \theta_g| \leq 1$, $\frac{\lambda}{d} \leq 1.8667$ or $d \geq 0.536\lambda$. Thus, if the inter-element spacing is greater than 0.536 times the wavelength, grating lobes could occur within $-90° < \theta_g < 90°$ region when the main beam is steered towards $\theta_{0(\mathrm{max})}$. Some of the energy in the grating lobe can spill into the operating cell sector, even if the maxima of the grating lobe occurs outside the $[\ -60°\quad 60°\ ]$ region.

The 3G UTRA spectrum allocations provide a nominal 190MHz separation between the up-link (mobile to basestation transmissions) and down-link

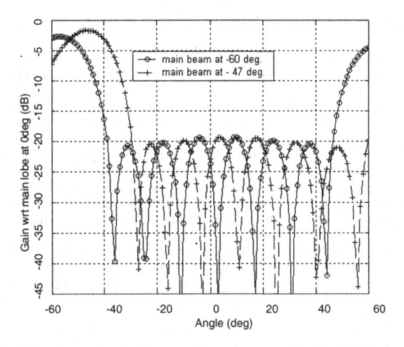

Fig. 2.6 Radiation patterns from the eight-element array showing the presence of a grating lobe at 78.8°.

(base station to mobile) frequencies. Analysing the frequency allocation in the UK [16] for example, it is evident that if $\lambda/2$ inter-element spacing is employed on the up-link carrier, grating lobes will occur in the down-link operation. Alternatively, selecting the down-link frequency for element spacing will reduce the electrical size of the array in the up-link, thus widening the main beam on the uplink frequency.

Figure 2.6 [14] shows the radiation pattern of an eight-element array with 0.546λ spacing between elements. When the main beam is steered to $-60°$ a grating lobe is shown to appear at 78.8°, and when the main beam is steered to $-47°$ the grating lobe is shown to have diminished. Note that amplitude weights derived from using the Dolph-Chebyshev technique have been incorporated, to achieve a 20 dB sidelobe level. Dolph-Chebyshev weighting is described in the following section. This example illustrates the impact of operating bandwidth on beamforming, where the radiation pattern is shown to change over the operating bandwidth. Chapter 3 discusses techniques for broadband beamforming which enable the radiation pattern to be controlled over a large range of frequencies (i.e. FB$\geq$ 1%). This is achieved by employing more complex weighting techniques that take into account some of the frequency dependent variations.

2.5 AMPLITUDE WEIGHTS

Consider a four-element array with $\lambda/2$ spacing between element and zero phase weights and unity amplitude weights applied. As the amplitude weighting of the two outside elements is reduced towards zero, the main beam width increases. This is illustrated in figure 2.7 where amplitude weights of $[1, 1, 1, 1]$ (crossed markers), $[0.5, 1, 1, 0.5]$ (dot markers), and $[0.05, 1, 1, 0.05]$ (dashed line) have been used. The radiation pattern of a two-element array with unity amplitude weighting is also plotted (solid line) and it can be seem that the radiation pattern of the four-element array approaches that of the two-element array as the amplitude weights of the outside elements are reduced. This also has the effect of reducing the sidelobe levels where the sidelobe level is <-40 dB when the outer element amplitude weightings are reduced to 0.05 (not visible in the figure). Note that the difference shown in the figure between the amplitude weighted four-element array pattern and the two-element array pattern is in the amplitude of the main beam, where the four-element array will transmit more power, i.e., through the two additional elements. This additional 'gain' can be eliminated by appropriate normalisation to allow direct comparison. Hence, with normalisation little difference is observed between the two main beams.

Window functions enable the amplitude weights to be controlled with certain constraints. The constraints relate to the characteristics of the choice of function, such as desired sidelobe level or main beam width, or a combination of both. These functions are commonly employed in other areas of signal processing such as temporal filter design and several types of window functions are investigated in the following section.

2.5.1 Window Functions

Sidelobes can be controlled to some extent by employing window functions. These are a family of functions that set the amplitude weights of the beamformer. Common window functions are: rectangular, Hamming, Hanning, Bartlett, Triangular, Kaiser and Dolph-Chebyshev, which have many applications in signal processing as well as beamforming. Each of these window functions are now investigated and the trade-offs between them observed in terms of main beamwidth (defined as the *half-power beamwidth* or HPBW) and sidelobe level (SLL). An eight-element uniform linear array with $d = \frac{\lambda}{2}$ is considered for each of the examples. The resulting characteristics are summarised in table 2.1. For each example a table of the amplitude weights is given which enable the characteristic taper of the weights that results from using a window function to be observed, with the peak occurring at the centre of the array and the minimum at each end. Note that the relationship between the array weights and the radiation pattern is through the Fourier transform, in the same manner as time and frequency are related in time domain signal processing. Thus, the radiation patterns shown

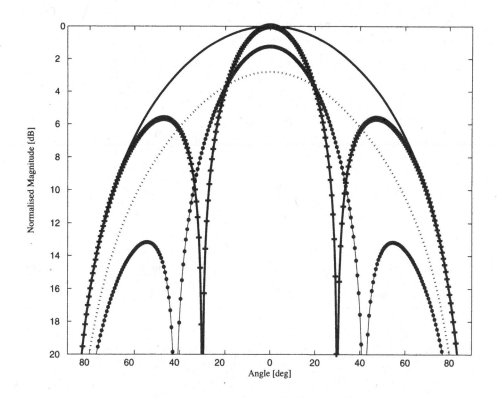

Fig. 2.7 Array pattern of a four-element ULA with various amplitude weights. The impact of amplitude weighting on the sidelobe level and main beam width can be seen. Array pattern of a two-element array with unity amplitude weights is shown for comparison (solid line).

for each of the following examples have been produced by plotting the Fourier transform of the weight (amplitude) vector computed from the respective window function. The weight vectors for each of the example window functions are shown in table 2.2. Note that the plots have been normalised so that the peak of the radiation pattern is at 0 dB.

2.5.1.1 Rectangular window. The rectangular window [17] gives uniform weighting to the array and the response is the equivalent of not applying a window. It gives the minimum mainlobe width at the expense of a relatively high sidelobe level. The amplitude weights are computed by equation (2.34).

$$W(n) = 1 \qquad (2.34)$$

The resulting radiation pattern is shown in figure 2.8(a).

Table 2.1 Half-power beamwidth and sidelobe levels resulting from various window functions applied to an eight-element ULA.

Window Function	HPBW	SLL [dB]
Rectangular	40°	−12.5
Bartlett	60°	−25
Blackman	80°	−50
Chebychev	36°	−20
Hamming	58°	−33
Hanning	50°	−32
Kaiser	36°	−15
Triangular	45°	−30

Table 2.2 Amplitude weights for various window functions for an eight-element array.

Element no.	1	2	3	4	5	6	7	8
Rectangular	1	1	1	1	1	1	1	1
Bartlett	0	0.2857	0.5714	0.8571	0.8571	0.5714	0.2857	0
Blackman	0	0.0905	0.4592	0.9204	0.9204	0.4592	0.0905	0
Triangular	0.1250	0.3750	0.6250	0.8750	0.8750	0.6250	0.3750	0.1250
Hamming	0.0800	0.2532	0.6424	0.9544	0.9544	0.6424	0.2532	0.0800
Hanning	0.1170	0.4132	0.7500	0.9698	0.9698	0.7500	0.4132	0.1170
Kaiser $\alpha=1$	0.7898	0.8896	0.9595	0.9595	0.9595	0.9595	0.8896	0.7898
Kaiser $\alpha=3$	0.2049	0.5010	0.7936	0.9754	0.9754	0.7936	0.5010	0.2049
Kaiser $\alpha=5$	0.0367	0.2707	0.6517	0.9552	0.9552	0.6517	0.2707	0.0367
D-Cheb. -20dB	0.578	0.659	0.8786	1	1	0.8786	0.659	0.578
D-Cheb. -30dB	0.2605	0.5176	0.8114	1	1	0.8114	0.5176	0.2605
D-Cheb. -40dB	0.1463	0.4182	0.7596	1	1	0.7596	0.4182	0.1463

2.5.1.2 Bartlett window.   The Bartlett window [17] is formed by the convolution of two rectangular windows. The characteristics are very similar to the triangular window except that, in this case, the computation of the weights results in zeros at the end elements. This has the effect of essentially reducing the array size and consequently broadens the main beam. The tapering of the amplitude weights of the remaining active elements reduces the sidelobes to a level below the level when no amplitude weighting is employed (i.e., a six-element array with rectangular weighting).   The amplitude weights with Bartlett windowing are computed from equations (2.35) and (2.36) depending on if the total number of elements is an odd or even number. The resulting radiation pattern is shown in figure 2.8(b).

For odd N:

$$W(n+1) = \begin{cases} \dfrac{2n}{N-1} & 0 \leq n \leq \dfrac{N-1}{2} \\[3mm] 2 - \dfrac{2n}{N-1} & \dfrac{N-1}{2} \leq n \leq N-1 \end{cases} \tag{2.35}$$

For even N:

$$W(n+1) = \begin{cases} \dfrac{2n}{N-1} & 0 \leq n \leq \dfrac{N}{2} - 1 \\[3mm] \dfrac{2(N-n-1)}{N-1} & \dfrac{N}{2} \leq n \leq N-1 \end{cases} \tag{2.36}$$

2.5.1.3 Blackman window. The Blackman window function [17] is an example of a generalised cosine window function consisting of a sum of sinusoids. In this case the function is a sum of three terms as shown below.

$$W(n) = 0.42 + 0.5 \times \cos\left(\frac{2\pi n}{N}\right) + 0.08 \times \cos\left(\frac{4\pi n}{N}\right) \tag{2.37}$$

The Blackman window yields very low sidelobes at the expense of a relatively wide main beam. The resulting radiation pattern is shown in figure 2.8(c).

2.5.1.4 Triangular window. The triangular window [17] is similar to the Bartlett window but offers a slightly improved main beam/sidelobe level trade-off. The amplitude weights computed when a triangular window is applied are computed using the equation below.

For odd N:

$$W(n) = \begin{cases} \dfrac{2n}{N+1} & 1 \leq n \leq \dfrac{N+1}{2} \\[3mm] \dfrac{2(N-n+1)}{N+1} & \dfrac{N+1}{2} \leq n \leq N \end{cases} \tag{2.38}$$

For even N:

$$W(n) = \begin{cases} \dfrac{2n-1}{N} & 1 \leq n \leq \dfrac{N}{2} \\[3mm] \dfrac{2(N-n+1)}{N} & \dfrac{N}{2} + 1 \leq n \leq N \end{cases} \tag{2.39}$$

The resulting radiation pattern is shown in figure 2.8(d).

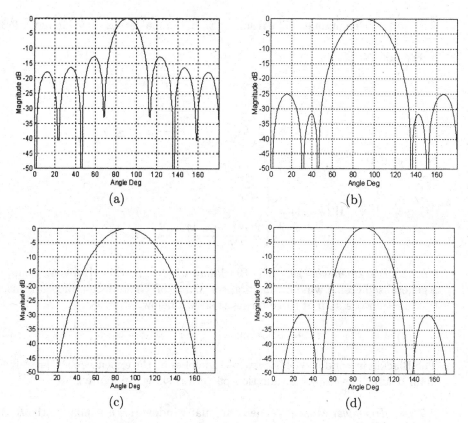

Fig. 2.8 Radiation pattern with: (a) Rectangular windowing, (b) Bartlett windowing, (c) Blackman windowing, (d) Triangular windowing.

2.5.1.5 *Hamming window.* The Hamming window [17] is another example of a generalised cosine window using two sinusoidal terms as shown below.

$$W(n+1) = 0.54 - 0.46 \times \cos\left(\frac{2\pi n}{N}\right) \qquad (2.40)$$

The resulting radiation pattern is shown in figure 2.9(a).

2.5.1.6 *Hanning window.* The Hanning window [17] is a further example of a generalised cosine window using two sinusoidal terms and is therefore similar to the Hamming window as shown below.

$$W(n+1) = 0.5 - 0.5 \times \cos\left(\frac{2\pi n}{N}\right) \qquad (2.41)$$

It yields a slightly narrower main beam width compared to that of the Hamming window whilst still maintaining a comparable sidelobe level. The resulting radiation pattern is shown in figure 2.9(b).

2.5.1.7 Kaiser window. The Kaiser window function [18] maximises the ratio of sidelobe energy to mainlobe energy. The parameter used to control the sidelobe level is α. The expression for computing the weights is given by equation (2.42), where I_o is a modified Bessel function of order zero.

$$W(n) = \frac{I_o(\alpha\sqrt{1 - (2(n - N/2))^2})}{I_o(\alpha)} \tag{2.42}$$

Figure 2.9(c) shows the radiation patterns for $\alpha = 1$, $\alpha = 2$ and $\alpha = 3$. It can be seen from the figure that large α gives lower sidelobe levels at the expense of mainlobe width and that for $\alpha = 1$, the performance is similar to that of a rectangular window.

2.5.1.8 Dolph-Chebyshev window. The Dolph-Chebyshev window function (often referred to as the *minmax window*) minimises the main lobe width for a specified sidelobe level thus giving control of the sidelobe level requirements. The sidelobes are equiripple as shown in figure 2.9(d).

This windowing method is more of a technique compared to simply computing the amplitude weights from a given function, which has been the case with the previous examples of window functions. It is an example of *array synthesis*, where the array pattern is designed for a given set of parameters, which in this case are the main beam width and sidelobe level. Amplitude weights using this technique can be computed as follows.

The radiation pattern of an array with an odd or even number of elements is a summation of cosine terms, which is achieved by simplifying equation (2.7). Thus, for an odd number of elements:

$$E(\theta, \phi) = \sum_{n=0}^{\left(\frac{N-1}{2}\right)} a_n \cos(2nu) \tag{2.43}$$

and for an even number of elements

$$E(\theta, \phi) = \sum_{n=0}^{\left(\frac{N-1}{2}\right)} a_n \cos\left[(2N + 1)nu\right] \tag{2.44}$$

where $u = \frac{\pi d}{\lambda} \sin\theta \sin\phi$, N is the number of elements, n is the nth element and a_n is the nth amplitude weight.

The Dolph-Chebyshev technique enables values of a_n to be computed for a given maximum sidelobe level. The sum of cosines in equations (2.43) and (2.44) consists

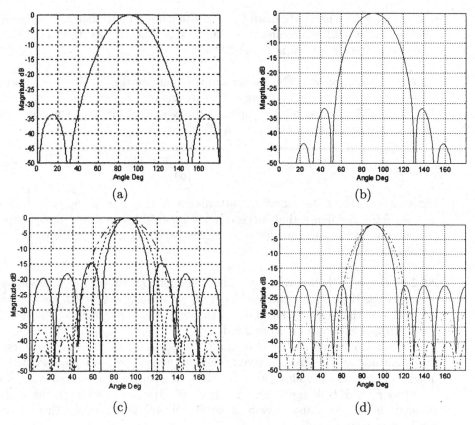

Fig. 2.9 Radiation pattern with (a) Hamming windowing (b) Hanning windowing (c) Kaiser windowing: solid line $\beta = 1$, dotted line $\beta = 3$, dashed line $\beta = 5$ (d) Chebychev windowing: solid line $R_s = -20$ dB, dotted line $R_s = -30$ dB, dashed line $R_s = -40$ dB.

of a fundamental term and $N - 1$ harmonics, as in Fourier analysis. Expanding these terms and finding an equivalent expression using Euler's formula

$$e^{jnm} = \cos(mn) + j\sin(mn) \tag{2.45}$$

yields the expressions given in table 2.3 for array sizes of $N = 1, \ldots, 8$. By letting $z = \cos(u)$, the expressions can be written in the form of Chebychev polynomials $(T_m(z))$ that are traditionally employed in filter design. These are shown in table 2.4. The polynomials in the table can be computed directly from

$$T_m(z) = \cos(m \cos^{-1} z) \qquad -1 \le z \le +1 \tag{2.46}$$

and

$$T_m(z) = \cosh(m \cosh^{-1} z) \qquad |z| > +1 \tag{2.47}$$

Table 2.3   Harmonics of array response.

N	Harmonic	$\cos(\mu)$	Equivalent expression
1	$m = 0$	$\cos(0)$	1
2	$m = 1$	$\cos(u)$	$\cos(u)$
3	$m = 2$	$\cos(2u)$	$2\cos^2(u) - 1$
4	$m = 3$	$\cos(3u)$	$4\cos^3(u) - 3\cos(u)$
5	$m = 4$	$\cos(4u)$	$8\cos^4(u) - 8\cos^2(u) + 1$
6	$m = 5$	$\cos(5u)$	$16\cos^5(u) - 20\cos^3(u) + 5\cos(u)$
7	$m = 6$	$\cos(6u)$	$32\cos^6(u) - 48\cos^4(u) + 18\cos^2(u) - 1$
8	$m = 7$	$\cos(7u)$	$64\cos^7(u) - 112\cos^5(u) + 56\cos^3(u) - 7\cos(u)$

Table 2.4   Chebychev polynomials.

N	Polynomial	$T_m(z)$
1	1	$T_0(z)$
2	z	$T_1(z)$
3	$2z^2 - 1$	$T_2(z)$
4	$4z^3 - 3z$	$T_3(z)$
5	$8z^4 - 8z^2$	$T_4(z)$
6	$16z^5 - 20z^3 + 5z$	$T_5(z)$
7	$32z^6 - 48z^4 + 18z^2 - 1$	$T_6(z)$
8	$64z^7 - 112z^5 + 56z^3 - 7z$	$T_7(z)$

The limit $-1 \le z \le +1$ represents the region of the sidelobes of the radiation pattern, and the region $|z| > +1$ represents the region of the main beam.

Using the example on an eight-element linear array with a required sidelobe level of -20 dB, an order 7 polynomial is required, i.e., $N = 7$. The sidelobe specification requires the main beam height to be 10 times higher than the maximum sidelobe level, i.e., $E_0 = 10$. Let z_0 be the angle of the main beam, thus using equation

$$T_7(z_0) = \cosh(7\cosh^{-1} z_0) = 10 \tag{2.48}$$

and rearranging for z_0 yields $z_0 = 1.0928$.

Now, the location of the main beam equates to $z_0 = 1$, therefore the polynomials must be normalised by

$$z' = \frac{z}{1.0928} = \frac{\cos u}{1.0928} \tag{2.49}$$

and therefore z_0 is substituted by z' throughout.

Expanding equation (2.43) for even elements gives

$$E(\theta, \phi) = a_3 \cos(7u) + a_2 \cos(5u) + a_1 \cos(3u) + a_0 \cos(u) \qquad (2.50)$$

Letting

$$E(\theta, \phi) = T_7(z) \qquad (2.51)$$

then

$$a_3 \cos(7u) + a_2 \cos(5u) + a_1 \cos(3u) + a_0 \cos(u) = 64z^7 - 112z^5 + 56z^3 - 7z \qquad (2.52)$$

Equating the above in terms of a_n gives

$$T_7(z) = a_3 \cos 7u + a_2 \cos 5u + a_1 \cos 3u + a_0 \cos u$$

$$= a_3 \left(64 \left(\frac{z}{z_0} \right)^7 - 112 \left(\frac{z}{z_0} \right)^5 + 56 \left(\frac{z}{z_0} \right)^3 - 7 \left(\frac{z}{z_0} \right) \right)$$

$$+ a_2 \left(16 \left(\frac{z}{z_0} \right)^5 - 20 \left(\frac{z}{z_0} \right)^3 + 5 \left(\frac{z}{z_0} \right) \right)$$

$$+ a_1 \left(4 \left(\frac{z}{z_0} \right)^3 - 3 \left(\frac{z}{z_0} \right) \right)$$

$$= 64z^7 - 112z^5 + 56z^3 - 7z \qquad (2.53)$$

Finally, equating the above in terms of $\cdot^x$ gives

$$z^7 \qquad 64a_3 = 64z_0^7 \qquad a_3 = z_0^7 = 1.86$$

$$z^5 \qquad 16a_2 - 112a_3 = -112z_0^5 \qquad a_2 = \frac{112}{16} (a_3 - z_0^5) = 2.12$$

$$z^3 \qquad 4a_1 - 20a_2 + 56a_3 = 56z_0^3 \qquad a_1 = \frac{1}{4} (56z_0^3 - 56a_3 + 20a_2) = 2.81$$

$$z \qquad a_0 - 3a_1 + 5a_2 + 7a_3 = -7z_0 \qquad a_0 = 7a_3 - 5a_2 + 3a_1 - 7z_0 = 3.2 \quad (2.54)$$

Note that the above equation yields four amplitude weights, however, the weighing is symmetrical about the centre of the array, i.e., $\frac{a_3}{2} = b_3 = b_7 \ldots$, where the weights have been normalised by a factor of 2. The weights can then be further normalised by the maximum value to give the amplitude weights below.

$$b_n = [\ 0.578 \quad 0.659 \quad 0.8786 \quad 1 \quad 1 \quad 0.8786 \quad 0.659 \quad 0.578\ ] \qquad (2.55)$$

2.6 CHAPTER SUMMARY

This chapter has focused on narrowband beamforming and has shown techniques enabling beam patterns with certain characteristics to be designed. Although the discussion has focused on the ULA and uniform planar array, the theory has been generalised enabling it to be applied to non-uniform array structures including circular and spherical arrays. Specific techniques for beamforming with circular arrays is discussed in chapter 5. The functions of the phase and amplitude weights applied to linear and planar arrays have been demonstrated. Examples of phase and amplitude weightings have been given. Furthermore, the impact of element spacing on the generation of (unwanted) grating lobes has been shown. Following this, numerous window functions have been introduced and compared. Beam synthesis through the use of the Dolph-Chebyshev technique has been demonstrated, which enables beam patterns with constant sidelobe levels to be constructed.

2.7 PROBLEMS

Problem 1. State the impact of phase weights and amplitude weights to an array radiation pattern.

Problem 2. Derive equation (2.11).

Problem 3. Compute the required phase weights to steer the main beam of a three-element ULA with $d = \frac{\lambda}{3}$ to $-30°$.

Problem 4. Compute the phase weight matrix for a 2×2 uniform planar array with $d_x = d_y = \frac{\lambda}{2}$ to steer the main beam towards an incoming signal at $\theta = +10°, \phi = +40°$. State a possible application for such a configuration.

Problem 5. Repeat example 2.2 assuming the required signal arrives at $-20°$ and the interfering signal is at $+35°$.

Problem 6. Show that a two-element array with an inter-element spacing of $d = 2\lambda$ and the main beam steered to $0°$ has grating lobes at $\pm30°$. How would this influence the performance if it was deployed in a mobile communications system? What can be done to eliminate grating lobes?

Problem 7. Compute the amplitude weights for a Hamming window function applied to a three-element ULA.

Problem 8. Construct a complex weight vector using the phase and amplitude weights from questions 3 and 7 respectively and plot the resulting array pattern. Compare with that produced by the same array when unity amplitude and phase weights are applied.

Problem 9. A ULA with six isotropic elements and $d = \frac{\lambda}{2}$ is to be designed so that the main beam is 20 dB above the sidelobes. Use the Dolph-Chebyshev technique to determine the amplitude weighting for each element. What effect will the reduced sidelobe level have on the main beam?

3

Wideband Array Processing

3.1 INTRODUCTION

Array signal processing or beamforming involves the manipulation of signals induced on the elements of an array. In a conventional narrowband beamformer, the signals from each sensor element are multiplied by a complex weight and summed to form the array output. Such a beamformer has formed the discussion in chapter 2. As the signal bandwidth increases the performance of the narrowband beamformer starts to deteriorate because the phase weight provided for each element and a desired angle should, for many applications, be constant with frequency. However, without suitable compensation these parameters will change for different frequency components of the communication wave. For processing broadband signals a tapped delay line (TDL) can normally be used on each branch of the array. The TDL allows each element to have a phase response that varies with frequency, compensating for the fact that lower frequency signal components have less phase shift for a given propagation distance, whereas higher frequency signal components have greater phase shift as they travel the same length. This structure can be considered to be an equalizer which makes the response of the array the same across different frequencies. Inherent baseband signals such as audio and seismic signals are examples of wideband signals. Also, in sensor array processing applications such as spread spectrum communications and passive sonar, there is growing interest in the analysis of broadband sources and signal processing.

Adaptive Array Systems   B. Allen and M. Ghavami
© 2005 John Wiley & Sons, Ltd   ISBN 0-470-86189-4

Wideband array signal processing has now become a reality due to the continued evolution and improvement of computing power. Whereas formerly wideband arrays required switched analogue delay elements and attenuators, now the non-adaptive array-pattern synthesis can be performed by digital filtering of sampled data, with the usual gains in precision and flexibility.

This chapter presents the fundamentals of wideband beamforming and several related beamforming techniques.

3.2 BASIC CONCEPTS

Most of the smart antenna techniques proposed in the literature relate to narrow-band beamformers. The antenna spacing of narrowband arrays is usually half of the wavelength of the incoming signal which is assumed to have a *fractional bandwidth* (FB) of less than 1%. By definition, the FB of a signal is the ratio of the bandwidth to the centre frequency as follows:

$$\text{FB} = \frac{f_h - f_l}{(f_h + f_l)/2} \times 100 \qquad (3.1)$$

where f_h and f_l are the highest and the lowest frequency components of the signal, respectively. Wideband arrays are designed for FB of up to 25% and *ultra wideband* (UWB) arrays are proposed for FB of 25 to 200%. Wideband and UWB arrays use a constant antenna spacing for all frequency components of the arriving signals. The inter-element distance, d, is determined by the highest frequency of the input wave and for a uniform one-dimensional linear array, is given by

$$d = \frac{c}{2f_h} \qquad (3.2)$$

where c is the wave propagation speed which may either be the speed of light for an electromagnetic wave or the speed of transversal (sound) wave through a medium. Wideband antenna arrays use a combination of spatial filtering with temporal filtering. On each branch of the array, a filter allows each element to have a phase response that varies with frequency. As a result, the phase shifts due to higher and lower frequencies are equalized by temporal signal processing.

Figure 3.1 shows the general structure of a wideband array antenna system. The TDL network permits adjustment of gain and phase as desired at a number of frequencies over the band of interest. The far-field wideband signal is received by N antenna elements. Each element is connected to $M - 1$ delay lines with the time delay of T seconds. The delayed input signal of each element is then multiplied by a real weight W_{nm} where $1 \leq n \leq N$ and $1 \leq m \leq M$. If the input signals are denoted by $x_1(t), x_2(t), \ldots, x_N(t)$, the output signal, which is the sum of all

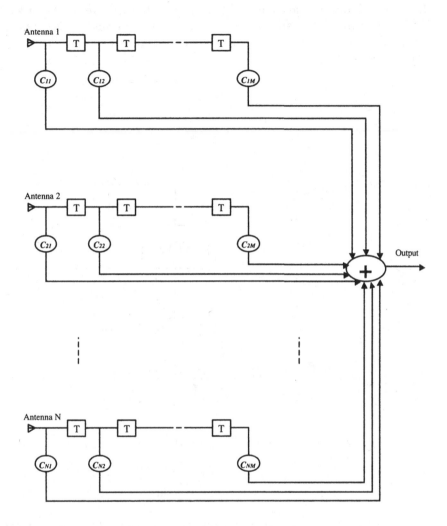

Fig. 3.1 General structure of a TDL wideband array antenna using N antenna elements.

intermediate signals, can be written as:

$$y(t) = \sum_{n=1}^{N} \sum_{m=1}^{M} W_{nm} x_n(t - (m-1)T) \tag{3.3}$$

In a linear array such as figure 3.1 [3], the signals $x_n(t)$ are related according to the angle of the arrival and the distance between the elements. Figure 3.2 shows that a time delay τ_n exists between the signal received at the element n and at the reference element $n = 1$. This amount of delay can be found as:

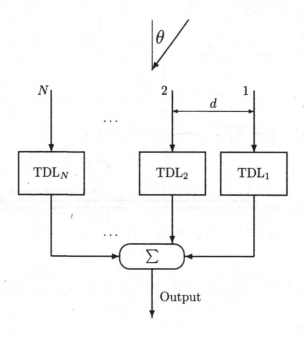

Fig. 3.2   Incoming signal is arriving at the antenna array under angle θ.

$$\tau_n = (n - 1)\frac{d}{c}\sin\theta \tag{3.4}$$

It is assumed that the incoming signal is spatially concentrated around the angle θ. Using the time delays corresponding to the antenna elements $x_n(t)$ can be expressed with respect to $x_1(t)$ as follows:

$$x_n(t) = x_1(t - \tau_n)$$
$$= x_1\left(t - (n - 1)\frac{d}{c}\sin\theta\right) \tag{3.5}$$

Different methods and structures can be utilised for computing the adjustable weights of a wideband beamforming network and some are explained in the following section.

3.3   A SIMPLE DELAY-LINE WIDEBAND ARRAY

The problem of designing a uniformly spaced array of sensors for narrowband operation is well understood. However, when a narrowband design is used over a wide bandwidth the array performance degrades significantly. At lower frequencies the beamwidth increases, resulting in reduced spatial resolution; at frequencies above the narrowband frequency the beamwidth decreases and grating lobes may be introduced into the array beam pattern as shown in figure 2.5. This is because the electrical distance between elements, which is measured in fractions of λ, becomes less for lower frequencies and more for higher frequencies. The consequence this has on the beam pattern can be determined from the associated analysis in chapter 2 and has the physical effects mentioned above. Additionally, this is why the inter-element spacing is better specified as a physical distance, d, instead of fractions (or multiples) of λ. Alternatively it can be specified as a function of a component signal frequency as given by equation (3.2).

Figure 3.3 shows the basic structure of a delay-line wideband transmitter array system. The adjustable delays T_n, $n = 1, 2, \ldots, N$, where N is the number of transducers, are controlled by the desired angle of the main lobe of the directional beam pattern, θ_0, as follows:

$$T_n = T_0 + (n - 1)\frac{d}{c}\sin\theta_0 \tag{3.6}$$

The constant delay $T_0 \geq (N - 1)d/c$ is required because without T_0, a negative delay will be obtained for negative values of θ_0 and cannot be implemented. The signal received at far field in the direction of θ, $(-90° < \theta < +90°)$, is equal to:

$$y(t) = A(\theta) \sum_{n=1}^{N} x_n(t - \tau_n)$$

$$= A(\theta) \sum_{n=1}^{N} x(t - T_n - \tau_n) \tag{3.7}$$

where $x_n(t)$ indicates the transmitted signal from transducer n, τ_n is the delay due to the different path lengths between the elements and the audience, and $A(\theta)$ is the overall gain of the elements and the path. The time delay τ_n from figure 3.3 is equal to

$$\tau_n = \tau_0 - (n - 1)\frac{d}{c}\sin\theta \tag{3.8}$$

where τ_0 is the constant transmission delay of the first element and is independent of θ. The gain $A(\theta)$ can be decomposed into two components as follows:

$$A(\theta) = A_1(\theta)A_2 \tag{3.9}$$

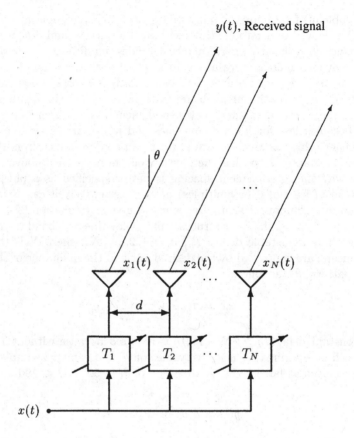

Fig. 3.3 Beam formation using adjustable delay lines.

where $A_1(\theta)$ is the angle-dependent gain of the elements and A_2 is the attenuation due to the distance. Substituting from equations (3.8) and (3.9) into equation (3.7) yields

$$y(t) = A_1(\theta)A_2 \sum_{n=1}^{N} x\left(t - \alpha_0 - (n-1)\frac{d}{c}(\sin\theta_0 - \sin\theta)\right) \qquad (3.10)$$

where $\alpha_0 = T_0 + \tau_0$. In the frequency domain, we may write

$$H(f,\theta) = \frac{Y(f,\theta)}{X(f)}$$

$$= A_1(\theta)A_2 e^{-j2\pi f\alpha_0} \sum_{n=1}^{N} e^{-j2\pi f(n-1)\frac{d}{c}(\sin\theta_0 - \sin\theta)}$$

$$= A_1(\theta)A_2 e^{-j2\pi f\alpha_0} e^{-j\pi f(N-1)\frac{d}{c}(\sin\theta_0 - \sin\theta)} \;.$$

$$\frac{\sin\left[\pi f N \frac{d}{c}(\sin\theta_0 - \sin\theta)\right]}{\sin\left[\pi f \frac{d}{c}(\sin\theta_0 - \sin\theta)\right]} \tag{3.11}$$

This equation enables several properties of a wideband delay beamformer to be derived. The directional patterns for different frequencies are the important characteristics of the beamformer.

Example 3.1

Consider an AM signal with the carrier frequency of 60 kHz and bandwidth of 10 kHz. Find and sketch the normalized amplitude of equation (3.11) for $\theta_0 = 10°$, $-90° < \theta < +90°$, $N = 10$, $d = 2.6$ mm, $c = 340$ m/s, and perfect transducers, i.e., $A(\theta) = A_2$. Note that this example is for acoustic (longitudinal) wave propagation through air.

Solution

The result is plotted in figure 3.4. We observe that at $\theta = \theta_0$ the frequency independence is perfect, but as we move away from this angle, the dependency increases. Nevertheless, the beamformer is considered wideband with a fractional bandwidth of $\frac{10}{60}$ or 17%. It should be noted that the same but unmodulated message (i.e., at baseband) would have a fractional bandwidth of $\frac{5}{2.5}$ or 200%. Increasing the inter-element spacing has positive and negative consequences. As will be seen shortly, it will produce a sharper beam and is more practical. On the other hand this increase will result in some extra main lobes in the same region of interest, i.e., $-90° < \theta < +90°$.

3.3.1 Angles of Grating Lobes

Now, using equation (3.11) the angles of the grating lobes are derived along with the condition for their existence. Assuming perfect transducers, i.e., $A(\theta) = A_2$, and, using equation (3.11) and for $\theta = \theta_0$

$$|y(f,\theta_0)| = A_2 N \tag{3.12}$$

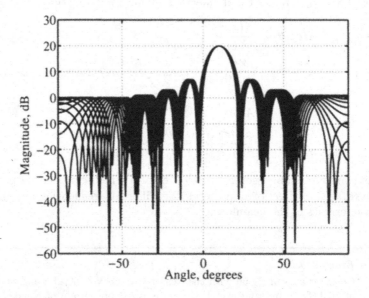

Fig. 3.4 Directional patterns of a delay beamformer for 11 frequencies uniformly distributed from 55 to 65 kHz.

This situation can happen for some other angles, denoted by θ_g. To calculate θ_g, it follows from equation (3.11) that

$$|H(f, \theta_g)| = A_2 N$$
$$= A_2 \frac{\sin\left[\pi f N \frac{d}{c}(\sin\theta_0 - \sin\theta_g)\right]}{\sin\left[\pi f \frac{d}{c}(\sin\theta_0 - \sin\theta_g)\right]} \qquad (3.13)$$

Now, equation (3.13) should be solved for θ_g,

$$\sin\left[\pi f \frac{d}{c}(\sin\theta_0 - \sin\theta_g)\right] = 0$$
$$\pi f \frac{d}{c}(\sin\theta_0 - \sin\theta_g) = m\pi \qquad (3.14)$$

where, $m = \pm 1, \pm 2, \ldots$. The result is

$$\theta_g = \sin^{-1}\left(\sin\theta_0 - m\frac{c}{df}\right) \qquad (3.15)$$

The first grating lobes are given for $m = \pm 1$. The condition of having no grating lobe for a beamforming network is that θ_g does not exist for any values of

$-90° < \theta_0 < +90°$. The worst case happens for $\theta_0 = \pm 90°$ and the condition of no grating lobe is written from equation (3.15) as follows

$$\frac{c}{df} \geq 2, \quad \text{or} \quad d \leq \frac{c}{2f} = \frac{\lambda}{2} \qquad (3.16)$$

where λ indicates the wavelength. It is interesting to note that θ_g is not a function of N, but is dependent on d. For more illustration, figure 3.4 is re-plotted for $d = 15$ mm in figure 3.5. It is observed that the nearest grating lobes for $f = 60$ kHz are at 33.5° and $-11.8°$ and are consistent with those obtained using equation (3.15). The frequency dependence of the beam patterns increases as we move away from the desired angle.

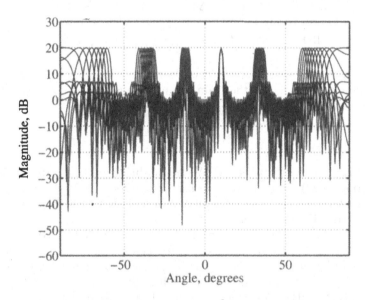

Fig. 3.5 Grating lobes have appeared as a result of the increase of spacing between the transducers.

3.3.2 Beam Width

Comparing figures 3.4 and 3.5 shows that the main beam width in figure 3.5 is less than that of figure 3.4. The corresponding equation that illustrates this property can be derived easily. The inter-null beamwidth (INBW) is defined as the difference between the nearest two nulls around the desired angle (often taken to be the centre of the main beam). Starting from equation (3.11) and equating it to zero gives the following:

$$\pi f N \frac{d}{c}(\sin\theta_0 - \sin\theta) = m\pi \qquad (3.17)$$

where, $m = \pm 1, \pm 2, \ldots$. The first two angles around θ_0 are denoted by θ_1 and θ_2, and are computed by rearranging equation (3.17) for $m = +1$ and $m = -1$, to give

$$\theta_1 = \sin^{-1}\left(\sin\theta_0 - \frac{c}{df N}\right) \qquad (3.18)$$

$$\theta_2 = \sin^{-1}\left(\sin\theta_0 + \frac{c}{df N}\right) \qquad (3.19)$$

respectively. Hence, the INBW $\Delta\theta = \theta_2 - \theta_1$ is written as

$$\text{INBW} = \sin^{-1}\left(\sin\theta_0 + \frac{c}{df N}\right) - \sin^{-1}\left(\sin\theta_0 - \frac{c}{df N}\right) \qquad (3.20)$$

From the previous analysis, it is clear that for $|\sin\theta_0 \pm \frac{c}{df N}| > 1$, there exists no null on either side of the main beam angle θ_0. Thus, for this special case, for $\theta_0 = 0$ gives

$$\text{INBW}_0 = 2\sin^{-1}\left(\frac{c}{df N}\right) \qquad (3.21)$$

i.e., increasing d lowers the INBW and produces sharper beams. It is easy to test equation (3.20) for values of the first and second cases which are illustrated in figures 3.4 and 3.5. The computed values of INBW for figures 3.4 and 3.5 are 25.6° and 4.4°, respectively.

It is obvious from equations (3.20) or (3.21) that INBW is a function of frequency f. To observe the effect of frequency variations on the beampattern of the delay-line beamforming network, we repeat example 3.1 for a wide frequency range from 45 kHz to 75 kHz. The result is given in figure 3.6 for spot frequencies of 45 kHz, 55 kHz, 65 kHz and 75 kHz. The computed values of INBW for these frequencies are 34°, 28°, 24° and 20°, respectively.

From the foregoing discussion we can conclude that pure delay-line wideband antenna arrays have the following properties:

- A relatively simple structure using only a variable delay element.

- Require no multiplier in the form of amplification or attenuation.

- A perfect frequency independence characteristic only for the desired main beam angle of the array.

- Their INBW and sidelobe characteristics vary considerably as a function of operating frequency.

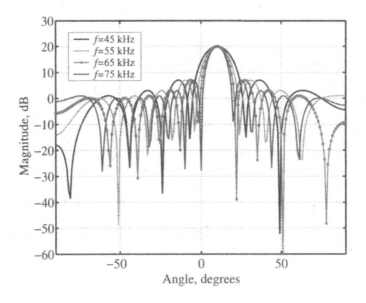

Fig. 3.6 Directional patterns of the delay beamformer at 45 kHz, 55 kHz, 65 kHz and 75 kHz.

3.4 RECTANGULAR ARRAYS AS WIDEBAND BEAMFORMERS

Rectangular linear arrays (i.e., uniform planar arrays), when subjected to an almost fixed elevation angle, may be used for fully spatial signal processing (i.e., in both azimuth and elevation) of signals with wideband properties [19]. In addition, the sharpness of the beams is maintained, not only at broadside, but also at the endfire directions of the array. The beamwidth of the directional pattern can be controlled at all angles and frequency domain filtering can be achieved easily in the design procedure. This can be called *frequency selective wideband beamforming* (FSWB) and compensates for the frequency dependence of the structure. Furthermore, we can selectively filter the wideband input signals in different frequency regions.

In this section, two algorithms are introduced for designing wideband rectangular antenna arrays. The first algorithm is based on the *inverse discrete Fourier transform* (IDFT) and the second algorithm is a constrained beamforming technique and includes matrix inversion. In both cases, the wideband signals received by the antenna are filtered in angle and frequency domains and interference coming from other sources with different *directions of arrival* (DOA) or frequencies are suppressed efficiently by spatial signal processing. The beamforming procedure is performed at *radio frequency* (RF) and real multipliers are necessary for each antenna element.

3.4.1   Rectangular Array Antenna in Azimuth

Figure 3.7 shows the structure of a uniform linear rectangular array configuration for $N_1 \times N_2$ elements of the proposed beamforming network. Each antenna element is denoted by (n_1, n_2), where $0 \leq n_1 \leq N_1 - 1$ and $0 \leq n_2 \leq N_2 - 1$ and has a frequency-dependent gain which is the same for all elements. The direction of the arriving signal is determined by the azimuth and the elevation angles θ and ϕ, respectively. As in most practical cases, it is assumed that the elevation angles of the incoming signals to the base station antenna array are almost constant and, without loss of generality, we consider $\phi \approx 90°$. The elements are placed at distances of d_1 and d_2 in the direction of n_1 and n_2, respectively. Assuming that the phase reference point is located at $(n_1 = 0, n_2 = 0)$, the phase of the incoming wave at the element denoted by (n_1, n_2) is

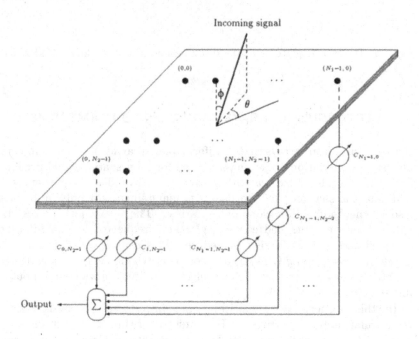

Fig. 3.7   Incoming signal is arriving at an $N_1 \times N_2$ array with azimuth angle θ and elevation angle ϕ. Each element is connected to a real multiplier, where not all of them are shown. Source: M. Ghavami, *Wide-band smart antenna theory using rectangular array structures*, IEEE Trans. Sig. Proc., Vol 50, No 9, pp 2143–2151, Sept, 2002 © IEEE.

$$\Phi(n_1, n_2) = \frac{2\pi f}{c}(d_1 n_1 \sin\theta - d_2 n_2 \cos\theta) \tag{3.22}$$

where f is the frequency variable.

Note that if ϕ was constant but not necessarily near $90°$, then we would have to modify d_1 and d_2 to new constant values of $(d_1 \sin\phi)$ and $(d_2 \sin\phi)$, respectively, which are in fact the effective array inter-element distances in an environment with almost fixed elevation angles of the incoming signals.

Assume that each antenna element is connected to one single coefficient $C_{n_1 n_2}$. Hence, the response of the array with respect to frequency and angle can be written as:

$$H(f,\theta) = G_a(f,\theta) \sum_{n_1=0}^{N_1-1} \sum_{n_2=0}^{N_2-1} C_{n_1 n_2} e^{j\frac{2\pi f}{c}(d_1 n_1 \sin\theta - d_2 n_2 \cos\theta)} \tag{3.23}$$

where $G_a(f,\theta)$ represents the frequency-angle dependent gain of each antenna element. Normally $G_a(f,\theta)$ is an even function of θ for $-90° < \theta < 90°$ and has a maximum value at $\theta = 0$, $f = f_0$, where $f_l < f_0 < f_h$.

Now, two auxiliary frequencies are defined by the following equations:

$$f_1 = \frac{fd_1}{c} \sin\theta \tag{3.24}$$

$$f_2 = \frac{fd_2}{c} \cos\theta \tag{3.25}$$

These new variables are functions of the frequency f and the angle θ, and are defined in this way in order to make the appearance of equation (3.23) as a two-dimensional (2D) discrete Fourier transform. They are related as follows:

$$\frac{f_1}{f_2} = \frac{d_1}{d_2} \tan\theta \tag{3.26}$$

This equation shows that, for any value of θ, if f is multiplied by a positive constant, then f_1 and f_2 will also be multiplied by the same factor and their ratio will be independent of frequency. This indicates a wideband property.

Substituting equations (3.24) and (3.25) in equation (3.23) yields

$$H(f_1, f_2) = G_a(f_1, f_2) \sum_{n_1=0}^{N_1-1} \sum_{n_2=0}^{N_2-1} C_{n_1 n_2} e^{j2\pi f_1 n_1} e^{-j2\pi f_2 n_2} \tag{3.27}$$

where $G_a(f,\theta)$ is now rewritten as a function of f_1 and f_2. The (2D) frequency region $f_1 - f_2$ is limited for both f_1 and f_2 to $(-0.5, 0.5)$, because, for example for f_1 we can write

$$|f_1| = \left|\frac{fd_1}{c} \sin\theta\right| \leq \frac{fd_1}{c} \leq \frac{f}{c}\frac{\lambda_{min}}{2} = \frac{f}{c}\frac{c}{2f_h} \leq 0.5 \tag{3.28}$$

where $\lambda_{\min}$ is the wavelength of the highest frequency. An expression similar to equation (3.28) is valid for f_2 as well.

A careful examination of equation (3.27) indicates that it looks like a *discrete Fourier transform* (DFT). Hence, the desired 2D frequency characteristics, divided by $G_a(f_1, f_2)$, gives the 2D DFT of the multipliers denoted by $C_{n_1 n_2}$. Therefore, by taking the IDFT of the values of $H(f_1, f_2)/G_a(f_1, f_2)$ in the $f_1 - f_2$ plane enables $C_{n_1 n_2}$ to be computed. This is the fundamental step of our first design method.

If φ is the polar angle of the $f_1 - f_2$ plane, we will have

$$\frac{f_1}{f_2} = \frac{d_1}{d_2} \tan \theta = \tan \varphi \tag{3.29}$$

and eliminating θ from equations (3.24) and (3.25) (which is achieved by squaring equations (3.24) and (3.25), finding $\sin^2$ and $\cos^2$ and adding them together knowing that $\sin^2 + \cos^2 = 1$) yields

$$\left(\frac{f_1}{fd_1/c} \right)^2 + \left(\frac{f_2}{fd_2/c} \right)^2 = 1 \tag{3.30}$$

which demonstrates an ellipse with the centre at $f_1 = f_2 = 0$. In the special case of $d_1 = d_2 = d$, we have circles with the equations $f_1^2 + f_2^2 = (fd/c)^2$ and radius fd/c. Equations (3.29) and (3.30) represent the loci of constant angle and constant frequency in the $f_1 - f_2$ plane, respectively. Plots of these two loci, given in figure 3.8, are helpful for determination of the angle and frequency characteristics of the wideband beamformer.

Assume that an array antenna system is to be designed with $\theta = \theta_0$, $f_l < f < f_h$ and a centre frequency of $f = f_0$. An example plot, showing the location of the desired points on the $f_1 - f_2$ plane is given in figure 3.9. This location is limited by $\varphi_0 = \tan^{-1}(\frac{d_1}{d_2} \tan \theta_0)$ and $r_l < |r| < r_h$, where,

$$r_l = \frac{f_l}{c} \bar{d} \tag{3.31}$$

$$r_h = \frac{f_h}{c} \bar{d} \tag{3.32}$$

and

$$\bar{d} = \sqrt{d_1^2 \sin^2 \theta_0 + d_2^2 \cos^2 \theta_0} \tag{3.33}$$

The symmetry of the loci with respect to the origin of the $f_1 - f_2$ plane results in real values for the multipliers' of each antenna element, $C_{n_1 n_2}$.

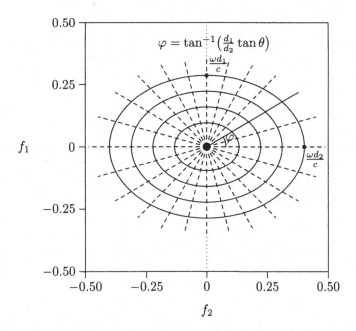

Fig. 3.8 The loci of constant angle θ and constant frequency f are radial and elliptical, respectively. Source: M. Ghavami, *Wide-band smart antenna theory using rectangular array structures*, IEEE Trans. Sig. Proc., Vol 50, No 9, pp 2143–2151, Sept, 2002 © IEEE.

3.4.2 Beamforming using IDFT

The first proposed design method of wideband beamforming with the specifications of figure 3.9, is explained in this subsection. In an ideal situation, i.e., for perfect omnidirectional antenna elements with $G_a(f_1, f_2) = 1$, the amplitude of $H(f_1, f_2)$ at the selected points given in figure 3.9 must be equal to one and at other points out of this region must be equal to zero. This is demonstrated by the following equation

$$
H_{\text{ideal}}(f_1, f_2) = \begin{cases} 1 & ; \quad \varphi_0 = \tan^{-1}(\frac{d_1}{d_2} \tan \theta_0) \ , \quad r_l < |r| < r_h \\ 0 & ; \qquad\qquad\quad \text{otherwise} \end{cases} \tag{3.34}
$$

where, r_l and r_h are as defined in equations (3.31) and (3.32), respectively. If the elements are not perfectly omnidirectional and all-pass, we have to compensate the frequency and angle dependence of the elements in equation (3.34). This can be

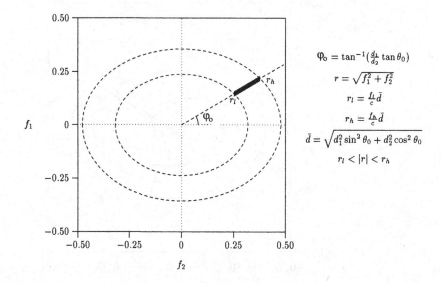

$$\varphi_0 = \tan^{-1}(\tfrac{d_1}{d_2}\tan\theta_0)$$
$$r = \sqrt{f_1^2 + f_2^2}$$
$$r_l = \tfrac{f_l}{c}\bar{d}$$
$$r_h = \tfrac{f_h}{c}\bar{d}$$
$$\bar{d} = \sqrt{d_1^2\sin^2\theta_0 + d_2^2\cos^2\theta_0}$$
$$r_l < |r| < r_h$$

Fig. 3.9 Location of the desired points on the intersection of the constant angle and constant frequency loci. Source: M. Ghavami, *Wide-band smart antenna theory using rectangular array structures*, IEEE Trans. Sig. Proc., Vol 50, No 9, pp 2143–2151, Sept, 2002 © IEEE.

achieved by replacing 1 in equation (3.34) with the inverse of $G_a(f_1, f_2)$ as follows:

$$\tilde{H}_{\text{ideal}}(f_1, f_2) = \begin{cases} G_a^{-1}(f_1, f_2) & ; \quad \varphi_0 = \tan^{-1}(\tfrac{d_1}{d_2}\tan\theta_0) \text{ and } r_l < |r| < r_h \\ 0 & ; \qquad\qquad\qquad \text{otherwise} \end{cases} \tag{3.35}$$

For designing the beamformer with beamwidth control, a low pass filter is defined by the following impulse response:

$$H_p(f) = \frac{\sin \pi f}{\pi f} \tag{3.36}$$

We then transform $H_p(f)$ to $H(f_1, f_2)$ as follows:

$$H(f_1, f_2) = \tilde{H}_{\text{ideal}}(f_1, f_2) H_p \left(\alpha \left(\frac{f_1}{f_2} - \frac{d_1}{d_2}\tan\theta_0 \right) \right) \tag{3.37}$$

where α is a constant, chosen according to the desired beamwidth of the antenna pattern, and θ_0 is the direction of the main lobe of the desired beamformer. Now, an IDFT is performed on the frequency response values of equation (3.37) and the results are $C_{n_1 n_2}$.

Now, several useful parameters of the wideband beamformer that can be selected adaptively are demonstrated.

3.4.2.1 Beamwidth.

To find a relation between INBW, $\Delta\theta$, and α, equations (3.36) and (3.37) are combined to give

$$H(f_1, f_2) = \tilde{H}_{\text{ideal}}(f_1, f_2) \frac{\sin\left[\alpha\pi\left(\frac{f_1}{f_2} - \frac{d_1}{d_2}\tan\theta_0\right)\right]}{\alpha\pi\left(\frac{f_1}{f_2} - \frac{d_1}{d_2}\tan\theta_0\right)} \tag{3.38}$$

Now, $H(f_1, f_2) = 0$ yields:

$$\alpha\left(\frac{f_1}{f_2} - \frac{d_1}{d_2}\tan\theta_0\right) = \pm 1, \pm 2, \ldots \tag{3.39}$$

The first zeros correspond to the values ± 1 in the right side of equation (3.39), which yields

$$\alpha\left(\frac{d_1}{d_2}\tan\theta - \frac{d_1}{d_2}\tan\theta_0\right) = \pm 1 \tag{3.40}$$

which can be rewritten as follows:

$$\theta = \tan^{-1}\left(\tan\theta_0 \pm \frac{1}{\alpha}\frac{d_2}{d_1}\right) = \theta_1, \theta_2 \tag{3.41}$$

This means that the nearest two zeros to θ_0 are θ_1 and θ_2, where $\theta_2 > \theta_1$. Therefore $\Delta\theta = \theta_2 - \theta_1$ can be written as

$$\Delta\theta = \tan^{-1}\left(\tan\theta_0 + \frac{1}{\alpha}\frac{d_2}{d_1}\right) - \tan^{-1}\left(\tan\theta_0 - \frac{1}{\alpha}\frac{d_2}{d_1}\right) \tag{3.42}$$

Assuming that θ_0 and $\Delta\theta$ are known, we can derive a relation for determining α from $\Delta\theta$ to be

$$\alpha = \frac{\frac{d_2}{d_1}\tan\Delta\theta}{-1 + \sqrt{1 + \tan^2\Delta\theta + \tan^2\Delta\theta\tan^2\theta_0}} \tag{3.43}$$

This relation is plotted in figure 3.10 for various values of θ_0 and $\Delta\theta$. As an example, for $d_1 = d_2$, $\theta_0 = 80°$ and $\Delta\theta = 20°$, we will have $\alpha = 0.275$, and the beamwidth is $20°$.

3.4.2.2 Number of antenna elements.

The dimensions of the rectangular array are determined by N_1 and N_2. The absolute value of the beamformer angle θ_0 can be related to the number of elements in the directions n_1 and n_2 in a way that for $\theta_0 = 0°$ and $|\theta_0| = 90°$, a maximum number of antenna elements in the directions

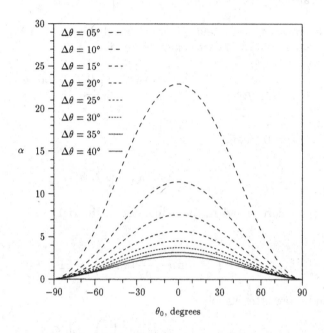

Fig. 3.10 Beamwidth parameter, α, as a function of the beam angle θ_0 and the beamwidth $\Delta\theta$. Source: M. Ghavami, *Wide-band smart antenna theory using rectangular array structures*, IEEE Trans. Sig. Proc., Vol 50, No 9, pp 2143–2151, Sept, 2002 © IEEE.

of n_1 and n_2, and a minimum number of elements in the directions of n_2 and n_1. Using a linear relationship

$$N_1 = \text{nint}\left[N_{1\text{max}} - \frac{|\theta_0|}{90^\circ}\left(N_{1\text{max}} - N_{1\text{min}}\right)\right] \tag{3.44}$$

and

$$N_2 = \text{nint}\left[N_{2\text{max}} - \frac{(90^\circ - |\theta_0|)}{90^\circ}\left(N_{2\text{max}} - N_{2\text{min}}\right)\right] \tag{3.45}$$

where 'nint' denotes the nearest integer to the argument. As an example, for $N_{1\text{max}} = N_{2\text{max}} = 41$, $N_{1\text{min}} = N_{2\text{min}} = 11$ and $\theta_0 = 0^\circ, 45^\circ, 90^\circ$, this gives $N_1 = 41, 26, 11$ and $N_2 = 11, 26, 41$, respectively. Therefore we can save on the number of elements used when designing different beamformers for various arrival angles.

The following illustrates the sharpness of the controlled beam pattern of the array at the endfire angles. The parameters of the array have already been mentioned above, except $f_l = 0.1c/d$ and $f_h = 0.48c/d$. Figure 3.11 shows the 2D frequency response and figure 3.12 demonstrates the beam pattern for 10 frequencies between f_l and f_h. The FB here is 131%.

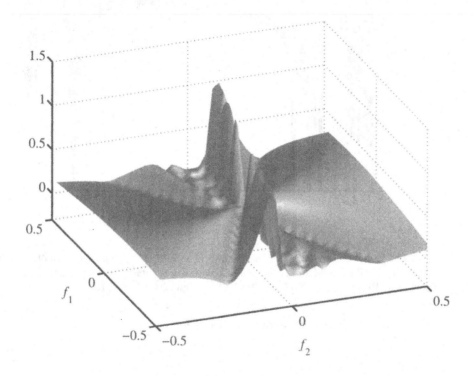

Fig. 3.11 2D frequency response of the IDFT method with $\theta_0 = 80°$. Source: M. Ghavami, *Wide-band smart antenna theory using rectangular array structures*, IEEE Trans. Sig. Proc., Vol 50, No 9, pp 2143–2151, Sept, 2002 © IEEE.

3.4.3 Beamforming using Matrix Inversion

In the second method (matrix inversion) for the design of the real multipliers, instead of controlling all points of the $f_1 - f_2$ plane, which is very difficult to do, we consider only L points on this plane. These L points are symmetrically distributed on the $f_1 - f_2$ plane and do not include the origin, thus L is considered

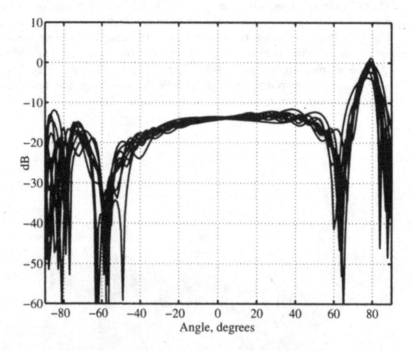

Fig. 3.12 Directional patterns of the beamformer for $f_l = 0.1\frac{c}{d}$ and $f_h = 0.48\frac{c}{d}$. Source: M. Ghavami, *Wide-band smart antenna theory using rectangular array structures*, IEEE Trans. Sig. Proc., Vol 50, No 9, pp 2143–2151, Sept, 2002 © IEEE.

an even integer. Two vectors are defined as follows:

$$\mathbf{b} = [b_1, b_2, \ldots, b_L]^T \tag{3.46}$$

$$\mathbf{H_0} = \left[H(f_{1_{0_1}}, f_{2_{0_1}}), H(f_{1_{0_2}}, f_{2_{0_2}}), \ldots, H(f_{1_{0_L}}, f_{2_{0_L}}) \right]^T \tag{3.47}$$

where the superscript ^{T} stands for the matrix transpose operation. The elements of the vector $\mathbf{H_0}$ have the same values for any two pairs $(f_{1_{0_l}}, f_{2_{0_l}})$, $l = 1, 2, \ldots, L$ which are located symmetrically with respect to the origin of the $f_1 - f_2$ plane. In addition, they consider the frequency dependence of the elements in a way similar to equation (3.35). The vector $\mathbf{b}$ is an auxiliary vector and will be computed in the design procedure. Now, assume that $H(f_1, f_2)$ is expressed by the multiplication of two basic polynomials and then the summation of the weighted result as follows:

$$H(f_1, f_2) = \sum_{l=1}^{L} b_l \left(\sum_{n_1=0}^{N_1-1} e^{j2\pi n_1(f_1 - f_{1_{0_l}})} \right) \left(\sum_{n_2=0}^{N_2-1} e^{-j2\pi n_2(f_2 - f_{2_{0_l}})} \right) \tag{3.48}$$

In fact with this form of $H(f_1, f_2)$, the problem of direct computation of $N_1 \times N_2$ coefficients $C_{n_1 n_2}$ from a complicated system of $N_1 \times N_2$ equations leads to a new problem of solving only L equations, because normally $L << N_1 \times N_2$. Nevertheless, the final task will be finding $C_{n_1 n_2}$ from b_l. The relationship between b_l and $C_{n_1 n_2}$ is obtained by rearranging equation (3.48) as follows:

$$H(f_1, f_2) = \sum_{n_1=0}^{N_1-1} \sum_{n_2=0}^{N_2-1} \left(\sum_{l=1}^{L} b_l e^{-j2\pi n_1 f_{1_{0_l}}} e^{j2\pi n_2 f_{2_{0_l}}} \right) e^{j2\pi n_1 f_1} e^{-j2\pi n_2 f_2} \quad (3.49)$$

Comparing with equation (3.27) yields

$$C_{n_1 n_2} = \sum_{l=1}^{L} G_a^{-1}(f_{1_{0_l}}, f_{2_{0_l}}) b_l e^{-j n_1 f_{1_{0_l}}} e^{j n_2 f_{2_{0_l}}} \quad (3.50)$$

i.e., after calculation of $\mathbf{b}$, $C_{n_1 n_2}$ can be found from equation (3.50). The computation of $\mathbf{b}$ is not difficult from equation (3.48). Define an $L \times L$ matrix $\mathbf{A}$ with the elements $\{a_{kl}\}$, $1 \leq k, l \leq L$ as follows:

$$a_{kl} = \sum_{n_1=0}^{N_1-1} e^{j2\pi n_1 (f_{1_{0_k}} - f_{1_{0_l}})} \sum_{n_2=0}^{N_2-1} e^{-j2\pi n_2 (f_{2_{0_k}} - f_{2_{0_l}})} \quad (3.51)$$

If equation (3.48) is rewritten for L pairs of $(f_{1_{0_l}}, f_{2_{0_l}})$, $l = 1, 2, \ldots, L$, then by using equations (3.46) and (3.47) it follows that:

$$\mathbf{H_0} = \mathbf{Ab} \quad (3.52)$$

Hence, $\mathbf{b}$ can be obtained from the following equation:

$$\mathbf{b} = \mathbf{A}^{-1} \mathbf{H_0} \quad (3.53)$$

It is assumed that $\mathbf{A}$ has a non-zero determinant, so that its inverse exists. Then, the values of $C_{n_1 n_2}$ are computed from equation (3.50). The question of how the symmetrical points $(f_{1_{0_l}}, f_{2_{0_l}})$ along with the corresponding gains of $H(f_{1_{0_l}}, f_{2_{0_l}})$ should be located in the $f_1 - f_2$ plane may arise. The following section considers the design algorithm for a 4×4 array with $L = 4$ selected points.

3.4.4 Numerical Examples

Two example simulations are given to illustrate performance based on the IDFT and matrix inversion techniques.

3.4.4.1 IDFT beamforming network. For this example of a frequency selective wideband beamformer network, it is assumed that a *software defined radio* (SDR)

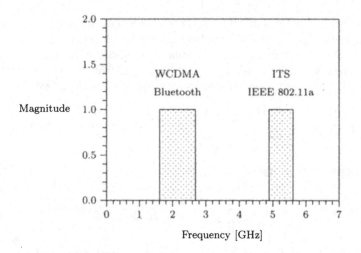

Fig. 3.13 The desired multi-bandpass behaviour of the beamformer for software-defined radio applications.   Source: M. Ghavami, *Wide-band smart antenna theory using rectangular array structures*, IEEE Trans. Sig. Proc., Vol 50, No 9, pp 2143–2151, Sept, 2002 © IEEE.

platform is designed for receiving RF signals with frequencies of 2.45 GHz and 5.25 GHz. It is desired that a single antenna system can perform simultaneous beamforming and frequency selection. We assume that a microstrip rectangular array with independent gain adjustment for $N_{1\max} = N_{2\max} = 41$ and $N_{1\min} = N_{2\min} = 11$ elements is available. Beamforming is done using fully analogue signal processing technology. The desired frequency response of the beamformer is shown in figure 3.13.

Each element of the antenna array has a frequency and angle response as follows:

$$G_a(f, \theta) = \left(-\frac{1}{80}f^2 + \frac{1}{20}f + \frac{19}{20}\right)\left(\frac{1}{16200}\theta^2 + 1\right) \qquad (3.54)$$

This approximate relation is valid for 1 GHz $< f <$ 6 GHz and $-90° < \theta < 90°$. The frequency f and the angle θ are measured in GHz and degrees, respectively. Note that equation (3.54) is a normalised equation and has a gain of unity $f = 2$ GHz and $\theta = 0°$. The bandpass behaviour of equation (3.54) in both frequency and angle makes it an appropriate measure of non-ideal properties of each antenna element. The antenna spacing is assumed to be $d_1 = d_2 = 0.025$ m. With substitution of $c = 3 \times 10^8$ m/s for the velocity of wave propagation, into equations (3.24) and (3.25) yields

$$f_1 = \frac{1}{12}f\sin\theta \qquad (3.55)$$

$$f_2 = \frac{1}{12} f \cos\theta \qquad (3.56)$$

where f is measured in GHz and f_1 and f_2 are normalised parameters without any unit. For obtaining the desired region of the $f_1 - f_2$ plane, it is necessary to calculate equations (3.29) and (3.30) as follows

$$\frac{f_1}{f_2} = \tan\theta = \tan\varphi \qquad (3.57)$$

$$f_1^2 + f_2^2 = \left(\frac{f}{12}\right)^2 \qquad (3.58)$$

Using these equations in equation (3.54), gives

$$G_a(f_1, f_2) = \left(-1.8f_1^2 - 1.8f_2^2 + 0.6\sqrt{f_1^2 + f_2^2} + 0.95\right) \cdot$$
$$\left(\frac{1}{16200}\left(\tan^{-1}\frac{f_1}{f_2}\right)^2 + 1\right) \qquad (3.59)$$

Since the maximum value of f_1 and f_2 is 0.5, the maximum operating frequency of the designed smart antenna will be 6 GHz. For each value of the beamformer angle, θ_0, it is desired that the frequency response of the beamformer has bandpass properties from about 1.6 GHz to 2.7 GHz and from about 4.9 GHz to 5.6 GHz and band rejection properties between these two ranges. Hence, with the consideration of a small certainty factor, the definition of the $H(f_1, f_2)$ is done according to the following equation

$$H(f_1, f_2) = \frac{\sin\left[\alpha\pi\left(\frac{f_1}{f_2} - \tan\theta_0\right)\right]}{\alpha\pi\left(\frac{f_1}{f_2} - \tan\theta_0\right)} \cdot$$

$$\begin{cases} \left[\left(-1.8f_1^2 - 1.8f_2^2 + 0.6\sqrt{f_1^2 + f_2^2} + 0.95\right)\left(\frac{1}{16200}\left(\tan^{-1}\frac{f_1}{f_2}\right)^2 + 1\right)\right]^{-1} \cdot \\ \qquad\qquad \text{for} \quad 0.38 < r < 0.49 \\ \qquad\qquad \text{and} \quad 0.079 < r < 0.24 \\ \\ \frac{1}{\sqrt{10}} \qquad\qquad\qquad\qquad \text{otherwise} \end{cases}$$

$$(3.60)$$

The value of θ_0 is set to 30° for this simulation. The parameter α is for controlling the bandwidth and is calculated to be 4.296, which corresponds to the beamwidth

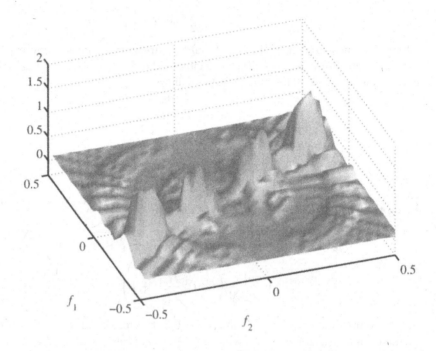

Fig. 3.14 A plot of $H(f_1, f_2)$ demonstrating the frequency selection and wideband properties of the proposed smart antenna. Source: M. Ghavami, *Wide-band smart antenna theory using rectangular array structures*, IEEE Trans. Sig. Proc., Vol 50, No 9, pp 2143–2151, Sept, 2002 © IEEE.

of 20°. Figure 3.14 demonstrates the magnitude of the $H(f_1, f_2)$ according to equation (3.60). Two desired regions in the lower and higher frequencies are emphasised by higher amplitudes. For a better presentation of the frequency selectivity of the beamforming network, in addition to the compensation of the imperfect antenna element specifications, the directional patterns of the system are plotted in figures 3.15–3.17. Figure 3.15 shows the directivity of the array for frequencies from 1.61 GHz to 2.69 GHz. Figure 3.16 illustrates the array patterns for the frequencies from 4.89 GHz to 5.61 GHz. In both figures, we observe no noticeable variation in the gains of the array antenna for the specified frequency bands.

In figure 3.17, the same patterns are plotted for frequencies from 3.49 GHz to 4.21 GHz. As expected, an attenuation of 10 dB, due to the factor $\sqrt{10}$ considered in equation (3.60), is observed for all related frequencies. Therefore the interference at the receiver is filtered in frequency as well as the angle (space) domains. The FB

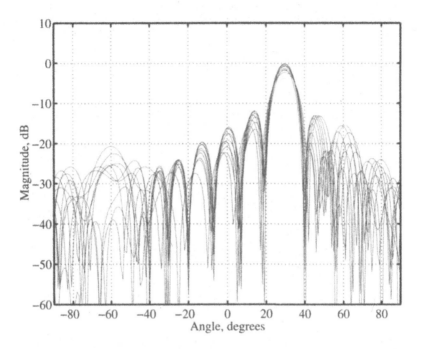

Fig. 3.15 Directional patterns of the beamformer plotted for frequencies from 1.61 GHz to 2.69 GHz. Source: M. Ghavami, *Wide-band smart antenna theory using rectangular array structures*, IEEE Trans. Sig. Proc., Vol 50, No 9, pp 2143–2151, Sept, 2002 © IEEE.

of the array is at least 0.14 and 0.5 for the higher and lower regions, respectively. Hence, the overall FB is more than 64%.

3.4.4.2 Matrix inversion beamforming network. In this example a simple 4×4 uniform linear rectangular array is designed and it is shown that the frequency independence of the array is appropriate for applications where the fractional bandwidth is supposed to be in the range of 0.2 to 0.3. The angle of the main beam is assumed to be $\theta_0 = -40°$ with the centre frequency of $f_0 = 0.35c/d$, where $d = d_1 = d_2$. Due to the limitation of the number of points on the $f_1 - f_2$ plane in this example, it is assumed that $G_a = 1$.

Initially, four pairs of critical points $(f_{1_{0_i}}, f_{2_{0_i}})$ are calculated as follows:

$$P_1: \quad (f_{1_{0_1}}, f_{2_{0_1}}) = (f_{1_0}, f_{2_0}) \tag{3.61}$$

$$P_2: \quad (f_{1_{0_2}}, f_{2_{0_2}}) = (-f_{1_0}, -f_{2_0}) \tag{3.62}$$

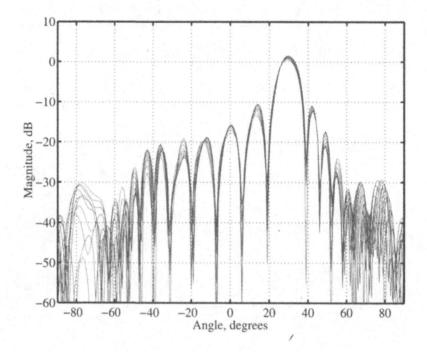

Fig. 3.16 Directional patterns of the beamformer plotted for frequencies from 4.89 GHz to 5.61 GHz. Source: M. Ghavami, *Wide-band smart antenna theory using rectangular array structures*, IEEE Trans. Sig. Proc., Vol 50, No 9, pp 2143–2151, Sept, 2002 © IEEE.

$$P_3 : \quad (f_{1_{0_3}}, f_{2_{0_3}}) = (-f_{2_0}, f_{1_0}) \tag{3.63}$$

$$P_4 : \quad (f_{1_{0_4}}, f_{2_{0_4}}) = (f_{2_0}, -f_{1_0}) \tag{3.64}$$

where f_{1_0} and f_{2_0} have been found from equations (3.24) and (3.25), respectively. Then, the vector $\mathbf{H_0}$ is formed as

$$\mathbf{H_0} = [1, 1, 0, 0]^T \tag{3.65}$$

Next, $\mathbf{A}$ is constructed using equation (3.51) and $\mathbf{b}$ is calculated from equation (3.53). Finally, $C_{n_1 n_2}$ for $1 \leq n_1, n_2 \leq 4$ is computed from equation (3.50). Due to the symmetry of the selected points in the $f_1 - f_2$ plane, the values of $C_{n_1 n_2}$ are all real. This simplifies the implementation.

The plot of $H(f_1, f_2)$ from equation (3.27), which is illustrated in figure 3.18, shows the actual 2D frequency response, obtained from the design algorithm. Clearly, there are two peaks at P_1 and P_2, and two zeros at P_3 and P_4. The important result of this pattern is that in a relatively large neighbourhood of the

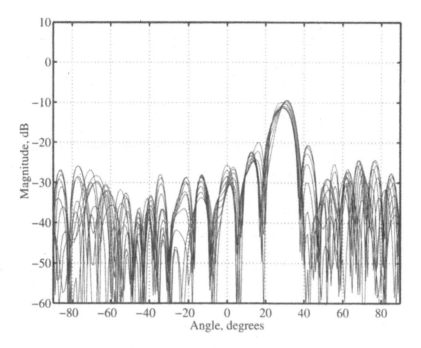

Fig. 3.17   Directional patterns of the beamformer plotted for frequencies from 3.49 GHz to 4.21 GHz. Source: M. Ghavami, *Wide-band smart antenna theory using rectangular array structures*, IEEE Trans. Sig. Proc., Vol 50, No 9, pp 2143–2151, Sept, 2002 © IEEE.

point corresponding to $f = f_0$, an almost constant amplitude is observed. For a 4×4 rectangular array this observation makes it a wideband array when it is designed for the centre frequency of the frequency band. Figure 3.19 demonstrates this fact more clearly. According to this figure, the frequency response for mainlobe and sidelobe invariance is from $f_l = 0.3c/d$ to $f_h = 0.4c/d$, i.e., a fractional bandwidth of 28.5%. Assuming a system with the carrier frequency of about 2.1 GHz, i.e., $f_0 = 2.1$ GHz, we will have

$$d = 0.35 \frac{c}{f_0}$$
$$= 0.05 \text{ m} \tag{3.66}$$

and the higher and lower frequencies will be $f_h = 2.4$ GHz and $f_l = 1.8$ GHz, respectively.

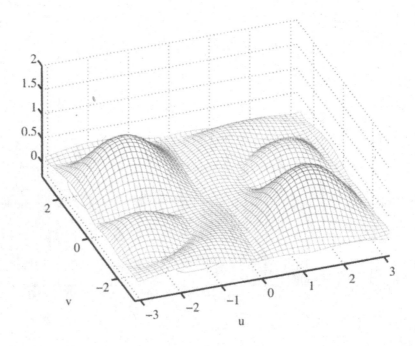

Fig. 3.18 The pattern of the 2D frequency response $H(f_1, f_2)$ for the desired 4×4 rectangular array.   Source: M. Ghavami, *Wide-band smart antenna theory using rectangular array structures*, IEEE Trans. Sig. Proc., Vol 50, No 9, pp 2143–2151, Sept, 2002 © IEEE.

3.4.5 Summary of Wideband Frequency Selective Rectangular Arrays

The following points regarding wideband frequency selective rectangular arrays can be considered:

- They are spatial signal processors for RF beamforming and do not require any time domain processing or complex multiplier phase shifting of the signals.

- Filtering of the incoming wave is done in the frequency and space domains without adding extra filters after beamforming.

- The wideband property of the antenna array was retained in spite of the frequency selection, and high values of FB could be obtained by different methods.

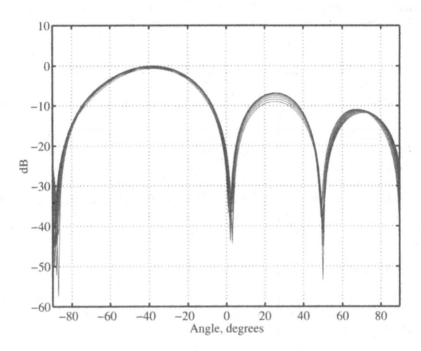

Fig. 3.19 Directional patterns of the 4 × 4 rectangular array for 10 equally spaced frequencies between $\omega_l = 0.3c/d$ and $\omega_h = 0.4c/d$. Source: M. Ghavami, *Wide-band smart antenna theory using rectangular array structures*, IEEE Trans. Sig. Proc., Vol 50, No 9, pp 2143–2151, Sept, 2002 © IEEE.

- Some imperfect behaviour of the elements, regarding frequency and angle, can be compensated for in the design. This can be helpful in practical design conditions. Furthermore, reception of a wide range of frequencies is done with a single antenna element spacing that is calculated for the highest frequency component of the input signal.

- Beamwidth and the number of elements selected in each direction of the array are calculated according to the analytical results. For the IDFT method, the sharp endfire beams are created with the desired beamwidth.

- It is predicted that using microstrip antenna array technology and analogue amplifiers this type of smart antenna can be very practical and cost effective.

Between the two discussed algorithms for computation of the real weights, the first method using IDFT, gave a more controlled beam pattern compared to the second method using matrix inversion. Instead, the second method, though for a larger beamwidth, required a lower number of antenna elements.

3.5 WIDEBAND BEAMFORMING USING FIR FILTERS

This section provides the basics of a general theory and design method for wideband arrays. It starts with development of the properties of a frequency invariant beam pattern for a sensor that is theoretically continuous, i.e., the aperture antenna, which differs from an array of elements that has formed the basis of the discussion so far. Note that an array of antenna elements is simply a sampled aperture. Thus the continuous sensor is then shown to be approximated by an array of discrete elements. The problem of designing a wideband array is then reduced to the problem of providing an approximation to a theoretically continuous sensor [20].

3.5.1 Continuous Linear Wideband Array

With regards to figure 3.20, the response of a linear continuous and one-dimensional aperture to planar waves from an angle θ measured relative to broadside can be written as

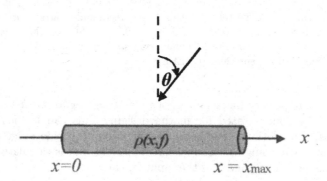

Fig. 3.20 A linear continuous and one-dimensional aperture.

$$h(\theta, f) = \int_0^\infty e^{j2\pi \frac{1}{c}xf \sin \theta} \rho(x, f) \, dx \qquad (3.67)$$

where $\rho(x, f)$ is a continuous function of both location x and frequency f, and is referred to as the *aperture illumination*, which shows the amplitude and phase of the field distribution over the antenna physical aperture. It is assumed that $\rho(x, f)$ is equal to zero for x greater than $x_{\max}$.

Assume that the aperture illumination of the sensor, $\rho(x, f)$, is given by

$$\rho(x, f) = fG(xf) \tag{3.68}$$

where $G(xf)$ is an arbitrary integrable complex function of a single variable. Then, the field pattern of equation (3.67) can be written as follows:

$$h(\theta, f) = \int_0^\infty e^{j2\pi \frac{1}{c}xf \sin \theta} fG(xf) \, dx \tag{3.69}$$

With the definition of a new variable $\zeta = xf$, this equation can be rewritten as

$$h(\theta, f) = \int_0^\infty e^{j2\pi \frac{1}{c}\zeta \sin \theta} G(\zeta) \, d\zeta$$
$$= h(\theta) \tag{3.70}$$

i.e., the beam pattern of the antenna is only a function of the angle. Hence, the filtering operation required for wideband beamforming can be broken into two parts:

- The primary filtering by

$$G_x(f) = G(xf) \tag{3.71}$$

- The secondary filtering, which is independent of position, by

$$\tilde{G}(f) = f \tag{3.72}$$

The important feature of the wideband antenna aperture is that all primary filters are related by dilation, i.e.,

$$G_{\alpha x}(f) = G(\alpha xf) = G_x(\alpha f) \tag{3.73}$$

3.5.2 Beamformer Implementation

The implementation of a wideband array is now described, where an array is defined as a practical structure that uses a finite set of identical, discrete, omnidirectional wideband sensors. The focus is on single-sided (i.e., an end element is considered as the reference instead of the centre element) one-dimensional array apertures with the first element located at $x = 0$.

An array of sensors can only approximate the ideal broadband continuous sensor. Assume that N sensor locations are identified by $x_0, x_1, \ldots, x_{N-1}$. The spacing between the array elements can be non-uniform. The frequency range of the design is limited to the interval $[f_l, f_h]$.

Now, the integral of equation (3.69) can be approximated by the following general equation

$$\tilde{h}(\theta, f) = f \sum_{n=0}^{N-1} g_n G(x_n f) e^{j2\pi \frac{1}{c} x_n f \sin \theta} \qquad (3.74)$$

where $f \in [f_l, f_h]$ and g_n is a frequency independent weighting function to compensate for the possibly non-uniform sensor locations. Using equation (3.71) the output of the primary filter attached to the nth sensor can be written as

$$Y_n(f) = G_{x_n}(f) e^{j2\pi \frac{1}{c} x_n f \sin \theta}$$
$$= G(x_n f) e^{j2\pi \frac{1}{c} x_n f \sin \theta} \qquad (3.75)$$

With regard to equation (3.73), and noting that $G_{x_n}(f) = G_{x_1}(\frac{x_n}{x_1} f)$, $Y_n(f)$ becomes

$$Y_n(f) = G_{x_1}\left(\frac{x_n}{x_1} f\right) e^{j2\pi \frac{1}{c} x_n f \sin \theta} \qquad (3.76)$$

This indicates that only one primary filter shape, i.e., G_{x_1}, is required in the numerical integration approximation.

As a special form of equation (3.74), the 'trapezoid' approximation to equation (3.69) can be written as [21]

$$\tilde{h}(\theta, f) = f \mathbf{Y}'(f) \mathbf{T} \mathbf{x} \qquad (3.77)$$

where ' $'$ ' stands for the matrix transpose operation,

$$\mathbf{Y}(f) = [Y_0(f), Y_1(f), \dots, Y_{N-1}(f)]' \qquad (3.78)$$

$$\mathbf{x} = [x_0, x_1, \dots, x_{N-1}]' \qquad (3.79)$$

and

$$\mathbf{T} = \begin{bmatrix} -0.5 & 0.5 & & & & & \\ -0.5 & 0 & 0.5 & & \bigcirc & & \\ & -0.5 & 0 & \ddots & & & \\ & & -0.5 & \ddots & 0.5 & & \\ & \bigcirc & & \ddots & 0 & 0.5 \\ & & & & -0.5 & 0.5 \end{bmatrix} \qquad (3.80)$$

In comparing the above trapezoid rule with the more general form of integration approximation in equation (3.74), the weighting functions g_n can be seen to relate to **Tx** via a non-trivial formula. For example, it can be shown that

$$g_0 = \frac{x_1 - x_0}{2}$$

$$g_1 = \frac{x_2 - x_0}{2}$$

$$\vdots$$

$$g_{N-1} = \frac{x_{N-1} - x_{N-2}}{2} \tag{3.81}$$

The weighting functions can be a function of one or more discrete sensor locations but are independent of frequency.

Figure 3.21 shows the simple format of the wideband array processor in a block diagram. This diagram demonstrates a number of important features:

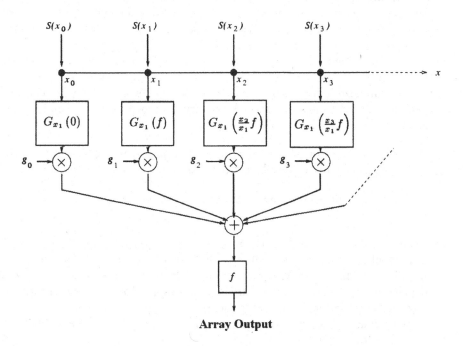

Fig. 3.21 Block diagram of a general one-dimensional wideband array with the origin at $x = 0$.

- The primary filters for all antenna elements are simple dilations of (i.e. derived from) a single frequency response.

- The primary filter outputs can be combined via frequency independent weights, g_n, that depend only on the sensor locations.

- All elements of the array share a common secondary filtering response, f, to generate the final output.

3.5.3   Sensor Locations

Sensor locations for the array implementation should be determined with the minimum number of elements required while maintaining performance. The major factor determining the minimum number of sensors possible is *spatial aliasing* which produces grating lobes if it is not suitably accounted for.

From the theory of linear uniformly spaced delay arrays and equation (3.16) it is well understood that grating lobes (i.e., periodic repetitions of the main beam) are introduced into the array beam pattern of an array if the spacing of array elements is greater than half of the wavelength of operation. This is referred to as spatial aliasing and is demonstrated for narrowband arrays in chapter 2 and wideband arrays earlier in this chapter.

Since the wideband aperture size scales with frequency, it is understood that the aperture size is constant if defined in terms of wavelength. It is assumed that the aperture size is finite and thereby defined as being P half-wavelengths at all frequencies, where, without loss of generality, P is restricted to be an integer. This highlights two related points:

(i)   since the aperture shape determines the primary filter shape then this implies the primary filter must be strictly bandlimited; and

(ii) for all frequencies except at the lowest design frequency some of the sensors are not used.

When the response of a sensor is used, i.e., the frequency of the signal lies in the primary filter passband, we will say the sensor is *active* at that frequency. In the following discussion we are referring only to active sensors. The locations of inactive sensors for a given frequency, despite the potential property that they violate a $\lambda/2$ spacing requirement, are completely irrelevant. Assume the desired frequency range is $[f_l; f_h]$. The finite aperture constraint implies a sensor positioning constraint

$$x_n = P\frac{\lambda_n}{2} \tag{3.82}$$

where n is the index of the active sensor of greatest distance from the origin, and λ_n is the wavelength corresponding to the bandwidth of the nth primary filter (or the highest frequency at which the nth sensor remains active).

The condition for a maximum spacing of $\lambda/2$ for all active sensors defines a second sensor positioning constraint:

$$x_n = x_{n-1} + \frac{\lambda_n}{2} \tag{3.83}$$

for $n > 0$, where n corresponds to the same condition as for the first sensor positioning constraint.

Combining these two constraints in equations (3.82) and (3.83) gives

$$x_n = \left(\frac{P}{P-1}\right) x_{n-1} \tag{3.84}$$

whenever $x_{n-1} > 0$, where P is the aperture size measured in half-wavelengths. This constraint must be maintained within the desired frequency range to avoid spatial aliasing. Since spacings less than $\lambda_h/2$, where λ_h is the wavelength corresponding to f_h, will not cause grating lobes at any frequency within the design band, it follows that the spacing within the densest portion of the array should be $\lambda_h/2$ to minimise the number of sensors. This densely packed portion of the array should have a total size of $P\lambda_h/2$ and will contain a minimum of $P+1$ sensors. Hence, the maximum spacing can be summarised as

$$x_n = \begin{cases} \dfrac{\lambda_h}{2} n, & \text{for } 0 \le n \le P \\[2ex] P\dfrac{\lambda_h}{2} \left(\dfrac{P}{P-1}\right)^{n-P}, & \text{for } P < n < N-1 \\[2ex] P\dfrac{\lambda_h}{2}, & \text{for } n = N-1 \end{cases} \tag{3.85}$$

where λ_l and λ_h are the wavelengths corresponding to the lower and upper design frequencies, respectively, P is the aperture size measured in half wavelengths, and N is the number of array elements. The maximum allowable spacing to avoid spatial aliasing, as defined by equation (3.85), is illustrated in figure 3.22. In the sense of producing an approximate broadband array which avoids spatial aliasing, this spacing relation represents the optimal sensor positioning function.

Using this optimal spacing relation, the minimum number of sensors required to implement a broadband array over a desired frequency range is

$$N = (P+1) + \left\lceil \frac{\log\left(\frac{f_h}{f_l}\right)}{\log\left(\frac{P}{P-1}\right)} \right\rceil \tag{3.86}$$

where $\lceil \cdot \rceil$ denotes the ceiling function.

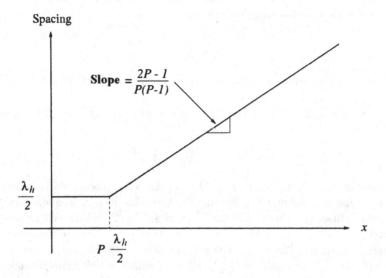

Fig. 3.22 Maximum permissible spacing of a single-sided one-dimensional array to avoid grating lobes.

3.5.4 Design of Primary Filters

The most important property of the primary filters of the wideband array is frequency dilation which means that all primary filters are derived from a single reference frequency response, and hence all primary filter coefficients may be derived from a single set of coefficients.

Two different methods of implementing finite impulse response (FIR) methods for a wideband beamformer are discussed.

3.5.4.1 Multi-rate method. The desired primary filter frequency response at a reference location x_r with a sampling period T is denoted by $G_{x_r}(f)$ and the corresponding filter coefficients are $g_{x_r}(k)$. The number of filter coefficients is L. The primary filters will have the required dilation property if the nth primary filter response is given by

$$G_{x_n}(f) = \sum_{k=-\frac{L-1}{2}}^{\frac{L-1}{2}} g_{x_r}(k)e^{-j2\pi f T_n k} \tag{3.87}$$

where $T_n = T x_n / x_r$ is the sampling period of the nth sensor.

A common method of multi-rate sampling is to sample every sensor at the highest rate required and then to use decimation to achieve the desired sampling rate. Thus, each of the primary filters would be implemented by down sampling by $\gamma_n = x_n/x_r$, applying the reference primary filter, and then up-sampling by γ_n. Note that γ_n must be an integer.

The aperture length is defined to be P half wavelengths at all frequencies within the desired band. Thus the nth primary filter is (ideally) bandlimited with

$$G_{x_n}(f) = 0 \qquad \text{for } |f| > \frac{Pc}{2x_n} \tag{3.88}$$

Ignoring the zeroth primary filter (which has a constant response) the primary filter with the widest bandwidth is located at

$$x_1 = \frac{c}{2f_h} \tag{3.89}$$

Hence the effective bandwidth is Pf_h, requiring a sampling rate of $f_s = 2Pf_h$. The reference primary filter is located at $x_r = c/(2f_h)$.

3.5.4.2 Single-rate method. As in the multi-rate case, assume there is a set of reference coefficients having a desired primary filter response at some reference location. By reconstructing the continuous time impulse response of $g_{x_r}(k)$, applying the scaling property of the Fourier transform, and resampling the scaled impulse response, the nth set of primary filter coefficients is given by

$$g_{x_m}(k) = \frac{1}{\gamma_n} \sum_{k=-\frac{L-1}{2}}^{\frac{L-1}{2}} g_{x_r}(k) \, \text{sinc}\left(\frac{m}{\gamma_n} - k\right) \tag{3.90}$$

where

$$\gamma_n = \frac{x_n}{x_r} \geq 1 \tag{3.91}$$

and

$$\text{sinc}(x) = \frac{\sin \pi x}{\pi x} \tag{3.92}$$

For $0 < \gamma_n < 1$ the reference set of coefficients must first be convolved with the coefficients of a lowpass filter having a cut-off of $\gamma_n f_s/2$ to avoid temporal aliasing. For $\gamma_n = 0$ the reference coefficients are simply an impulse. The length of the nth primary filter should be $L\gamma_n$ for $\gamma > 1$; a predefined number of coefficients should be used for $0 < \gamma_n < 1$ to give best results.

If the input signal is bandlimited to f_h, the minimum sampling rate is $f_s = 2f_h$. Recall that (ideally) $g_{x_n}(f) = 0$ for $|f| > Pc/(2x_n)$. The location of the reference sensor should be chosen such that $g_{x_r}(f) = 0$ for $|f| > f_s/2$. Hence, for the single sampling rate wideband array the reference sensor is located at $x_r = Pc/(2f_h)$.

3.5.4.3 Determining the coefficients of the reference primary filter. With the substitution $\rho(x, f) = fG(xf)$ and the change of variables $u = c^{-1}\sin\theta$ and $y = xf$ in equation (3.67), the Fourier transform relationship between the desired response and the aperture distribution is apparent. Thus, given a desired response $r_d(u)$ defined in the range $u \in [-1/c, 1/c]$, i.e., the visible region, the required aperture distribution is

$$G(y) = \int_{-\frac{1}{c}}^{\frac{1}{c}} r_d(u)e^{-j2\pi yu}\,du \tag{3.93}$$

Defining $\tilde{H}_x(f) = G(xf)$ as the primary filter response of the sensor at x, it is clear that

$$\tilde{H}_1(f) = \int_{-\frac{1}{c}}^{\frac{1}{c}} r_d(u)e^{-j2\pi fu}\,du \tag{3.94}$$

and that the impulse response of the primary filter at $x = 1$ is identical to the desired frequency independent beam pattern response, $r_d(\cdot)$. The set of coefficients for this primary filter can then be found by direct sampling as $h_1[k] = r_d(kT)$, where T is the sampling period. By the scaling property of the Fourier transform, the coefficients of the primary filter at some location x_r are given by

$$h_r[k] = \frac{1}{x_r}r_d\left(\frac{kT}{x_r}\right) \tag{3.95}$$

where $x_r > 0$. Note that since $u \in [-1/c, 1/c]$, the length of the reference filter is limited.

3.5.5   Design of Secondary Filters

Since the secondary filter is a differentiator, its design is straightforward. The secondary filter should be a Type 4 FIR filter, i.e., even length with odd symmetric coefficients.

3.5.6   Numerical Examples

Two examples are presented to demonstrate the effectiveness of each of the methods. In both examples the aperture length is $P = 4$ half wavelengths and the design bandwidth was 200–3400 Hz, requiring 17 sensors and a total array size of 3.4 m (for acoustic waves propagating in air). A secondary filter with 12 coefficients and a uniform aperture illumination is used in both cases.

Figure 3.23 shows the response of the multi-rate wideband array with a maximum sampling rate of 30 kHz and a reference filter with nine coefficients. Figure 3.24 shows the response of the single sampling rate FIB with a sampling rate of 8 kHz. The reference filter again had nine coefficients, but each of the primary filters has a minimum of 51 coefficients (with the largest primary filter having 151 coefficients). It is seen that in both cases the beam pattern is relatively frequency invariant over the entire design bandwidth. The multi-rate method requires fewer filter coefficients, but at the expense of a higher sampling rate.

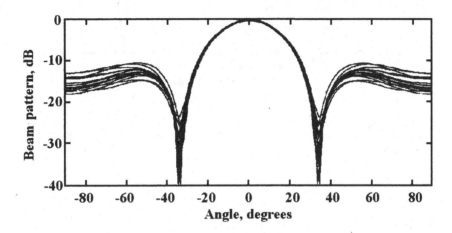

Fig. 3.23 Beam pattern of a 17-element wideband array at 15 frequencies within the design bandwidth using multi-rate processing.

3.6 CHAPTER SUMMARY

This chapter introduced the wideband array processing in contrast to the conventional narrowband techniques. Delay-line array systems and related topics such as angles of grating lobes and inter-null beamwidth were explained and derived. It was shown that rectangular linear arrays may be used for fully spatial signal processing of antenna arrays with wideband properties. Two different types of algorithms for rectangular wideband structures were investigated.

Finally, the basic theory of wideband beamforming using FIR filters was introduced. Different algorithms for implementation of the method and sensor location were discussed.

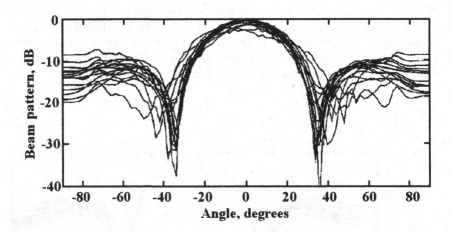

Fig. 3.24 Beam pattern of a 17-element wideband array at 15 frequencies within the design bandwidth using a single sampling rate.

3.7 PROBLEMS

Problem 1. In *Example 3.1* consider

$$d = \frac{4c}{f_h} \tag{3.96}$$

(i) Find d.

(ii) Calculate and sketch the inter-null beamwidth as a function of frequency f, $f_l < f < f_h$.

(iii) Compute the first two grating lobes at each side of the main lobe for $f = 50, 60, 70$ kHz.

(iv) Sketch the beam patterns of the array for 21 frequencies equally distanced from 50 to 70 kHz.

Problem 2. What is the effect of elevation angle on the rectangular wideband arrays?

Problem 3. Show that the auxiliary frequencies f_1 and f_2 are limited to ± 0.5.

Problem 4. Compare two algorithms of the rectangular wideband arrays with respect to:

(i) sharpness of the beams.

(ii) complexity.

(iii) number of antenna elements.

Problem 5. Derive equation (3.81).

Problem 6. Compare multi-rate and single-rate methods for wideband beamforming using FIR filters.

4

Adaptive Arrays

4.1 INTRODUCTION

Adaptive arrays are able to dynamically update their weights to respond to
changing signal conditions. Thus the weights are usually computed according
to the characteristics of the received signals, which are periodically sampled. The
ability to self-update is extremely desirable in many applications where the signals
change, such as in a mobile communications system, radar target tracking etc.

This chapter presents a toolbox of techniques that form the constituent parts of
an adaptive array. The choice of parts is determined by the design requirements
such as signal characteristics, technology and complexity and it is up to the system
designer to make these choices.

Figure 4.1 shows the classification of the techniques presented and discussed
in this chapter. Many of the algorithms require the computation of the *spatial
covariance matrix*, and is therefore introduced at the beginning of the chapter.
This is followed by multi-beam and scanning arrays which are particularly useful
for radar and sonar applications. The remainder of the chapter focuses on: switch
beam; temporal and spatial reference techniques; and blind beamforming. Switch
beam beamformers are considered to be partially adaptive since the beam cannot
be adjusted to the same extent as fully adaptive systems but do have several
advantages associated with them. Blind beamforming does not require the pilot
signals of other techniques that are used for channel estimation. Instead they
employ other estimation techniques and consequently are more computationally
intensive.

Adaptive Array Systems   B. Allen and M. Ghavami
© 2005 John Wiley & Sons, Ltd   ISBN 0-470-86189-4

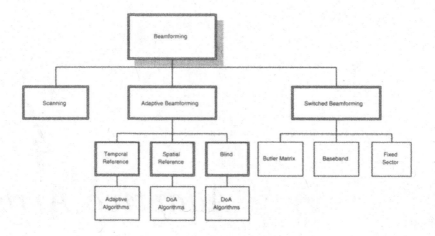

Fig. 4.1   Beamformer Classification.

4.2   SPATIAL COVARIANCE MATRIX

The spatial covariance matrix, $\mathbf{R_u}$, forms a fundamental part of many beamforming and direction-of-arrival algorithms. The performance of these algorithms is therefore dependent upon the estimation of $\mathbf{R_u}$ and due to the importance of this parameter, it is described here to enable the understanding of the forthcoming algorithms to be more accessible. This section describes how $\mathbf{R_u}$ can be estimated and how it impacts the convergence time of the adaptive beamformer. The subscript 'u' refers to the covariance matrix being computed on the *uplink* (mobile to base station) of a wireless system, i.e., the base station where the array is located is in receive mode. Conversely $\mathbf{R_d}$ refers to the *downlink* spatial covariance matrix, i.e., the basestation is transmitting. The significance of $\mathbf{R_u}$ and $\mathbf{R_d}$ is discussed in chapter 5 where downlink (transmit) beamforming is addressed.

The spatial covariance matrix, $\mathbf{R_u}$, is given by equation (4.1) where E[·] signifies the expectation and $\mathbf{u}$ is the received signal vector observed across the array.

$$\mathbf{R_u} = \mathbf{E[uu^H]} \qquad (4.1)$$

$\mathbf{R_u}$ can be estimated as follows. The structure of $\mathbf{R_u}$ is shown by equation (4.2) for an N element array. The first diagonal represents the auto-correlation of the received signal vectors observed at each element.

$$\mathbf{R_u} = \begin{bmatrix} \mathbf{R_{11}} & \mathbf{R_{11}} & \cdots & \mathbf{R_{1N}} \\ \mathbf{R_{21}} & \ddots & & \vdots \\ \vdots & & \ddots & \vdots \\ \mathbf{R_{N1}} & \cdots & \cdots & \mathbf{R_{NN}} \end{bmatrix} \tag{4.2}$$

Where $\mathbf{R_{NN}}$ is the cross-covariance of element outputs NN assuming a zero mean process. $\mathbf{R_{NN}}$ can be estimated by equation (4.3) where $x(k)$ and $y(k)$ are the K received signals from elements x and y, respectively.

$$\mathbf{R_{xy}} = \frac{1}{K} \sum_{k=1}^{K} x(k) \cdot y(k) \tag{4.3}$$

It is evident that by increasing K, temporal averaging is applied to $\mathbf{R_{NN}}$, which in turn is used to determine the appropriate array weights. For adaptive algorithms, a number of estimations of $\mathbf{R_{NN}}$ are required before convergence is obtained and convergence typically occurs after 50 iterations [22]. The reference reports a typical value of $L = 50$ for *recursive least squares* (RLS) and *square route recursive least squares* (SQRLS) algorithms operating on an eight8-element array with $\lambda/2$ spacing in a microcellular environment. The RLS algorithm is described later in this chapter, and the SQRLS algorithm is an extension of this. Thus, the convergence time, C, can be estimated by equation (4.4) where K is the number of samples, L is the number of iterations required for convergence and T_s is the sampling period.

$$C_\tau = K \cdot L \cdot T_s \tag{4.4}$$

Example 4.1

According to equation (4.4), the convergence time is a function of the number of samples, number of iterations and the sampling period of the digital beamformer. Assume a third-generation mobile communications system base station utilises a sampling period of 130 ns (i.e., twice the specified chipping rate of 3.85 Mc/s, although many systems employ oversampling techniques) and that the RLS algorithm is used so 50 iterations are required to achieve convergence. The resulting convergence time is plotted in figure 4.2 as a function of the number of samples. It is seen from the figure that 1539 samples are required to converge within 10 ms, i.e., the specified frame period for such systems. The beamformer should converge within one frame, hence a maximum of 1539 samples can be obtained and used for averaging. This illustrates a particular problem when short packets are transmitted. Such transmissions are required to be short bursts of data, but successful transmission will have to wait for the system to converge.

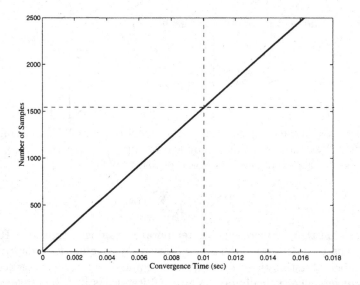

Fig. 4.2 Beamformer convergence characteristics for a 3G mobile communications system with sampling period of 130 ns and using the RLS algorithm.

4.3 MULTI-BEAM ARRAYS

The discussion so far in this book has focused on a beamforming system forming a single beam and controlling it accordingly. However, applications such as mobile communications systems or radar may require many users/targets to be monitored simultaneously. This can be done through a single array structure but with additional weight vectors, i.e., one per user as shown in figure 4.3. The figure shows a multiple-beam array with two beams directed, one directed at each of two users, where each user has an associated weight vector (beamformer), but a single array is used. Furthermore, each beam could have a different function, i.e., one tracking a user and the other searching (scanning) for new targets. Such a scenario is akin to radars, and scanning is described in the next section.

4.4 SCANNING ARRAYS

So far, the beamformers have been used to produce a constant beam shape in a particular direction. For many applications such an approach is appropriate, however, many applications require the beam to be scanned across the aperture. A radar system is such an application when it is searching for new targets. Once

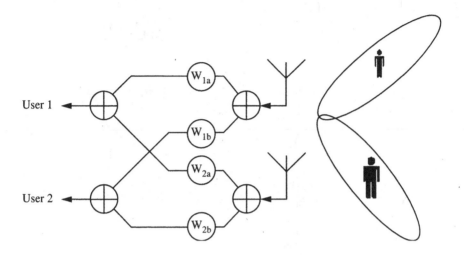

Fig. 4.3 A multi-beam array system.

the scanning beamformer has acquired a target it will be handed to a conventional beamformer in order to track the new target. Scanning will then resume.

Such a function is implemented by incrementing the weight vector, $\mathbf{W}$, to steer the angle of the main beam across the array aperture, which could be either azimuth or elevation scanning or in both dimensions as illustrated in figure 4.4. Thus $\mathbf{W}$ is updated at the desired scan rate. The weights could be computed from a single algorithm, or by sequentially reading the weights in from a memory device where they have been pre-computed and stored. Successive weights could increment the beam by either small or large angles, where small angular increments would result in a more continuous (higher resolution) scanning function.

Strictly speaking the scanning arrays described above are not classified as adaptive because they follow a programme sequence instead of dynamically adapting to a certain scenario. Thus they serve to illustrate the difference between adaptive arrays and other techniques. Another way of implementing a scanning array is to employ a feedback control system that determines a wait and hold function in a certain direction. Figure 4.4 illustrates the function of a scanning array where a beam is shown to scan in both azimuth and elevation.

4.5 SWITCHED BEAM BEAMFORMERS

Switched beam systems are technologically the simplest, and can be implemented using a number of fixed, independent, directional antennas, or virtually, with an

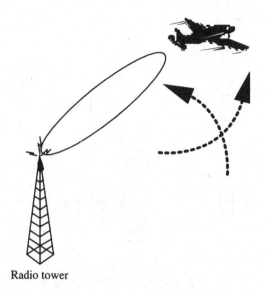

Radio tower

Fig. 4.4 A scanning array system.

antenna array and analogue beamformer such as a Butler matrix [23, 24, 25] or
Rotman lens [26]. Figure 4.5 shows the construction of a four-port Butler matrix,
which consists of four hybrid couplers, shown in figure 4.5(b). The system is easily
expanded to support eight beams by repeating the apparent pattern for a further
four ports. The operation of a Butler matrix can be likened to a fast Fourier
transform and yields M mutually orthogonal beams with a 4 dB cusping loss, as
shown in figure 4.6. The orthogonality of the beams is defined by the angle minima
of one beam pattern corresponding with the main beam angle of all of the other
beams.

A Rotman lens [26] is an alternative implementation and is also based on an
N by N port system. The operation relies on wave propagation between the
input and output ports of electrical lengths giving the required phase shift. Due
to the physical size constraints, it is unsuitable for current second and third-
generation mobile wireless network frequency allocations, and is therefore best
suited to systems operating with much higher frequency allocations.

A similar technique called *grid of beams* (GOB) can be used with digital
beamforming systems, which selects the best weights from a stored set. This,
however, leads to a more complex implementation due to the drawbacks associated
with digital beamforming highlighted in chapter 5. GOB does not require
the beams to be orthogonal enabling the beam cusp to be adjusted and an
arbitrary number of beams can be formed, each with a 3 dB beamwidth given

by equation (4.5) (assuming uniform array weighting) [15]. The beam cusp is defined in figure 4.6. This loss affects users located at angles corresponding to the cross-over between beams since:

1. Signals will be attenuated by this loss.

2. It can be difficult to determine which beam to assign to these users. In this case hysteresis in the beam selection can help to reduce this problem.

$$\theta_{3dB} \approx \frac{0.886\lambda}{Md\cos(\theta_o)} \tag{4.5}$$

Where M = number of elements, d = element spacing, λ = free space wavelength and θ is the steering angle from boresight. Furthermore, applying a window function such as a Hanning or Hamming window can control the sidelobe levels as shown in chapter 2. A GOB beamformer was implemented in the TSUNAMI II field trials system, which used 11 beams distributed over a 120° sector [27]. Results showed this technique to be unsuitable in interference limited systems as the wanted and unwanted signal sources could not be identified when beams are selected on a best-received SNIR basis. This, however, excludes applications in this form to most wireless communication networks. An extension to this technique reported in [28] and [29] overcomes this by using training sequences or user colour codes to distinguish between users. [29] reports a carrier-to-interference ratio (CIR) improvement of between 1 dB and 4 dB using a four beam system with the best improvement occurring for high received signal-to-noise plus interference ratio (SNIR). Switched beam systems implemented using a Butler matrix or RF beamforming network also have the advantage of being able to form a direct replacement for existing spatial diversity systems. They can also offer computational stability in multipath environments and enhance services to network hot spots [30, 31].

An interesting phenomenon reported in [32] is the *cartwheel effect* which describes the non-uniform uplink CIR distribution throughout the cell when the antenna system is deployed in a cellular network. This is caused by radial sectors of low CIR that point directly at the first tier of co-channel base stations (i.e., neighbouring co-channel base stations). Two solutions are proposed to mitigate this effect: dynamic channel assignment and realigning the beams to redistribute the bad areas. The reference also reports on simulation results of a switched beam system in a TDMA network. The uplink results show a 2 dB coverage improvement with respect to a 120° sector, two-branch space diversity system.

Reference [31] reports on a switch beam trials system deployed in a dense urban environment. The four beam, 40° beamwidth system yielded a 3 dB CIR improvement over an omni-sector antenna. Here a non-orthogonal beam set was used to reduce the cusping loss by 3 dB and beams were selected on a best-received signal strength basis.

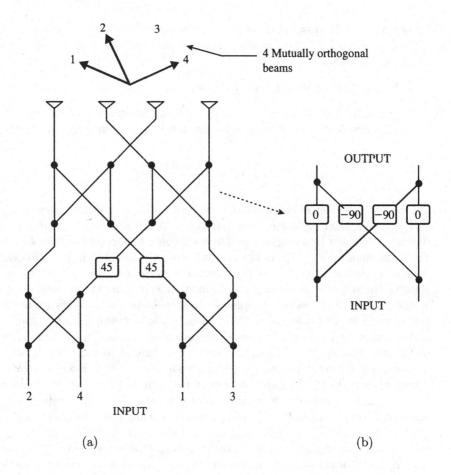

Fig. 4.5 (a) 4 × 4 Butler matrix. (b) A hybrid used to form a Butler matrix. Source: B. Allen, M. Beach, *On The Analysis of Switch-Beam Antennas for the W-CDMA Downlink*, IEEE Trans. Veh. Tech., Vol 53, No 3, pp 569–578, May, 2004 © IEEE.

4.6 FULLY ADAPTIVE BEAMFORMERS

Beamforming can be used to simultaneously receive a signal arriving from a several directions and attenuate signals from other directions. Systems designed to receive spatially propagating signals often encounter the presence of interfering signals. If the desired signal and interference occupy the same frequency band, unless the signals are uncorrelated, e.g., signals used in CDMA wireless systems, then temporal filtering often cannot be used to separate signals from interference. However, the desired and interfering signals often originate from different angles. This spatial separation can be exploited to separate signals from interference using

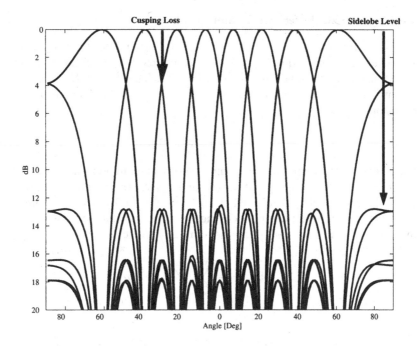

Fig. 4.6 Eight orthogonal radiation patterns formed by an eight port Butler matrix.

a spatial filter at the receiver. A beamformer is a processor used in conjunction with an array of sensors, i.e., antenna elements in an adaptive array, to provide a versatile form of spatial filtering as introduced in chapter 1. The array collects spatial samples of propagating wave fields, which are processed by the beamformer. Typically a beamformer linearly combines the spatially sampled time series from each sensor to obtain a scalar output time series in the same manner that an FIR filter linearly combines temporally sampled data. Furthermore, in the same manner that an adaptive FIR filter dynamically adjusts its weights to provide an optimum output, an adaptive array will adjust its weights to provide an optimum spatial filter. The weights for such a spatial filter can be computed in the time (*temporal reference beamforming*) or spatial (*spatial reference beamforming*) domain. They can also be computed without the use of a reference signal, which is referred to as *blind beamforming*. These three techniques are described in the remainder of this chapter.

4.6.1 Temporal Reference Beamforming

Temporal reference beamforming uses the temporal signal properties, such as embedded training sequences, to form a reference signal (this can be viewed as an extension to maximal ratio combining [33]). The Wiener-Hopf optimum weight vector [34], **Wu**, can then be computed from the temporal signal properties using equation (4.6).

$$\mathbf{W_u} = \mathbf{R_u}^{-1}\mathbf{p_u} \tag{4.6}$$

$\mathbf{R_u}$ is the $M \times M$ spatial covariance matrix computed from equation (4.7) and $\mathbf{p_u}$ is the $M \times 1$ cross-correlation vector between the antenna output, $x(n)$, and reference signal, $r(n)$ computed using equation (4.8). $(\cdot)^H$ signifies Hermitian transpose and $(\cdot)^*$ the complex conjugate. Computation of the spatial covariance matrix has been discussed at the beginning of this chapter.

$$\mathbf{R_u} = \frac{1}{N} \sum_{n=0}^{N-1} x(n) \cdot x^H(n) \tag{4.7}$$

$$\mathbf{p_u} = \frac{1}{N} \sum_{n=0}^{N-1} r(n) \cdot x^*(n) \tag{4.8}$$

The antenna weights are continuously updated to optimise a quantity such as: *minimum mean square error* (MMSE); *maximum signal to interference plus noise ratio* (MSINR) or *minimum variance distortionless response* (MVDR) (or minimum interference power). Asymptotically, all criteria lead to the same solution given by the Wiener-Hopf equation (equation (4.6)) [35]. A major advantage of temporal reference beamforming is the ability to maximise the output SNIR even if the number of interfering sources exceeds $M - 1$, where M is the number of antenna elements. An array with M elements has M *degrees of freedom* (DoF) where one DoF is required to track the desired user, thus leaving $M - 1$ available for controlling the remainder of the array pattern. In this case, the beamforming system will assign the available degrees of freedom to cancel the strongest interferers first and will then proceed to the weaker ones in turn until either is exhausted. Array calibration requirements are reduced here since any signal distortion caused by the antenna manifold can often be corrected for in the adaption process of the beamformer and does not alter the optimisation problem. This is one advantage of temporal reference beamforming since it enables 'self-correction'. Many other beamforming techniques require careful calibration to correct for hardware distortions. This is covered in detail in chapter 5. Note that temporal reference beamforming systems often benefit from calibration since it enables more rapid convergence to the optimum weight vector.

The performance of temporal reference beamformers depends on the adaptive algorithm that updates the beamformer weights. Typical adaptive algorithms for this application include *direct matrix inversion* (DMI) (also known as sample matrix inversion); *least mean squares* (LMS) and *recursive least squares* (RLS). References [33] and [34] provide a detailed discussion of the above algorithms together with a number of modifications.

A temporal reference beamformer was implemented as part of the TSUNAMI II project using an 8-element array [27]. The field trials results showed a mean C/I gain of 8.5 dB compared to a single element with 120° 3 dB beamwidth in a microcellular environment. Performance degradation was observed when the number of signals was incorrectly estimated and shows a limitation of the algorithm.

Another characteristic attributed to temporal reference beamformers is that of signal cancellation. This occurs when the desired signal is not accurately known or in low SNIR scenarios. In these circumstances, the poorly estimated signal vector is interpreted as interference. Two methods of mitigating this are: to extract only the noise plus interference spatial covariance matrix; or by diagonally loading the joint spatial covariance matrix. These are reported in [36] and [37], respectively.

4.6.2 Spatial Reference Beamforming

Spatial reference beamforming may not make use of embedded training sequences. Instead, the *directions of arrival* (DoA) of the impinging waves are used to synthesise beams steered at the wanted signal with nulls directed to other co-channel users (interference). The DoA characteristics are a function of the multipath environment which is discussed further in chapter 5.

The performance of DoA null steering methods depends mainly on two factors:

- Assumption that the spatial structure of the impinging signal from each source is accurately described by only one steering vector, which means that there is insignificant azimuth spreading of the signals, e.g., line-of-sight scenario.

- Performance of the DoA algorithm itself, and distortion caused by the antenna manifold.

Any variation in these factors will therefore cause performance degradation. Direction-finding methods require precise antenna manifold information, which necessitates fast correction of the calibration errors.

An integral part of a spatial reference beamformer is the DoA estimation algorithm. These algorithms vary in performance and complexity. These methods can be classified into *spectral* and *parametric* based techniques [38]. In the former, a spectrum-like function is computed, i.e., the DoA spectrum. The locations of the highest (separated) peaks are the desired DoA estimates. Typical methods in this group include (in ascending order of performance): classical *Fourier Method* (FM) (or Bartlett) [39], *Capon* [40] and *MUltiple SIgnal Classification* (MUSIC) [41]. While these methods are computationally attractive, they do not always provide the required accuracy, especially in the mobile radio environment with coherent multipath sources. If high accuracy is required, the alternatives are the parametric methods. The increased robustness and accuracy are obtained at the expense of complexity since multi-dimensional optimisation is usually required. If the multi-dimensional search is performed iteratively, an initial guess can be provided by the spectral methods. Typical methods in this group include: Deterministic and Stochastic *maximum likelihood* (ML) [42, 43], *estimation of signal parameters via rotational invariance techniques* (ESPRIT) [44], *space alternating generalised expectation maximisation* (SAGE) [45] and *weighted subspace fitting* (WSF) [46]. Table 4.1 summarises the advantages and disadvantages of candidate DoA techniques. Unitary ESPRIT [47] is a computationally efficient modification to the standard ESPRIT algorithm. An important requirement for these algorithms is real-time parameter estimation, making many of the super resolution algorithms inappropriate due to the high computational complexity creating an intolerable delay. The performance of spatial reference beamformers can be improved by utilising the received signal training sequence to discriminate between the wanted and undesired signals. However, their great advantage of reduced complexity over temporal reference algorithms is then lost.

4.7 ADAPTIVE ALGORITHMS

Most adaptive beamforming algorithms may be categorised into two classes according to whether a training signal is used or not. One class of these algorithms is non-blind adaptive algorithms in which a training signal is used to adjust the array weight vector. On the other hand, blind adaptive algorithms do not require a training signal. This class is discussed in the final section of this chapter.

In non-blind adaptive algorithms, a training signal, $d(t)$, which is known to both the transmitter and receiver, is sent from the transmitter to the receiver during the training period. The beamformer in the receiver uses the information of the training signal to compute the optimal weight vector, $\mathbf{W_{opt}}$. After the training period, data is sent and the beamformer uses the weight vector computed previously to process the received signal. If the channel and the interference characteristics remain constant from one training period to the next, the weight vector $\mathbf{W_{opt}}$ will remain unchanged. The training signal therefore samples the channel and must obey the well-known Nyquist sampling theorem. The remainder of this section reviews the development of well-known adaptive algorithms used in adaptive antennas. Initially the Wiener solution is developed, which subsequent adaptive algorithms are used to compute. The application of adaptive algorithms avoids direct computation of the Wiener solution which would otherwise require

Table 4.1 Comparison of direction-of-arrival estimation methods.

Algorithm	Advantages	Disadvantages
FM	Simple implementation, Robust to element perturbations	Depends on array size
Capon	Better resolution than FM	Limited by sensor noise power
MUSIC	Good resolution	Lower performance than ESPRIT Sensitive to gain and phase errors Sensitive to coherent multipath
ML	Statistically optimum results	Computationally complex Requires many snapshots
ESPRIT	High resolution, Non-critical array calibration	Computationally complex Limited by array geometries Requires multiple snapshots
WSF	Performance with coherent interferers	Limited by estimation accuracy

matrix inversion which is computationally intensive and can lead to an unstable result.

4.7.1 Wiener Solution

Most non-blind algorithms are designed to minimize the *mean-squared error* between the desired signal $d(t)$ and the array output $y(t)$. This is termed the *Wiener solution* and the following shows how it is developed for the signals relating to an adaptive beamformer. Let $y(k)$ and $d(k)$ denote the sampled signal of $y(t)$ and $d(t)$ at time instant t_k, respectively. Then the error signal is given by

$$e(k) = d(k) - y(k) \tag{4.9}$$

and the mean-squared error is defined by the cost function

$$J = E[|e(k)|^2] \tag{4.10}$$

where $E[\cdot]$ denotes the expectation value. It should be noted that the cost function J attains its minimum when all the elements of its gradient vector are simultaneously zero. Substituting equation (4.9) and $y(k) = \mathbf{W}^H\mathbf{x}(k)$ into equation (4.10) gives

$$J = E[|d(k) - y(k)|^2] \tag{4.11}$$
$$= E[\{d(k) - y(k)\}\{d(k) - y(k)\}^*] \tag{4.12}$$
$$= E[\{d(k) - \mathbf{W}^H\mathbf{x}(k)\}\{d(k) - \mathbf{W}^H\mathbf{x}(k)\}^*] \tag{4.13}$$
$$= E[|d(k)|^2 - d(k)\mathbf{x}^H(k)\mathbf{W} - \mathbf{W}^H\mathbf{x}(k)d^*(k) + \mathbf{W}^H\mathbf{x}(k)\mathbf{x}^H(k)\mathbf{W}] \tag{4.14}$$
$$= E[|d(k)|^2] - \mathbf{p_u}^H\mathbf{W} - \mathbf{W}^H\mathbf{p_u} + \mathbf{W}^H\mathbf{R_u}\mathbf{W} \tag{4.15}$$

where

$$\mathbf{R_u} = E[\mathbf{x}(k)\mathbf{x}^H(k)] \tag{4.16}$$

and

$$\mathbf{p_u} = E[\mathbf{x}(k)d^*(k)] \tag{4.17}$$

In equation (4.15), $\mathbf{R}$ is the $M \times M$ auto-correlation matrix of the input data vector $\mathbf{x}(k)$, and $\mathbf{p}_u$ is the $M \times 1$ cross-correlation vector between the input data vector and the desired signal $d(k)$. The gradient vector of $\mathbf{J}$, i.e. $\Delta(\mathbf{J})$ is defined by

$$\nabla(\mathbf{J}) = 2\frac{\partial \mathbf{J}}{\partial \mathbf{W}^*} \tag{4.18}$$

where $\frac{\partial}{\partial \mathbf{W}^*}$ denotes the conjugate derivative with respect to the complex vector $\mathbf{W}$.

When the mean-squared error $\mathbf{J}$ is minimised, the gradient vector will be equal to an $M \times 1$ null vector, i.e.,

$$\nabla(\mathbf{J})|_{\mathbf{W_{opt}}} = 0 \tag{4.19}$$

Substituting equation (4.15) into equation (4.18) gives

$$-2\mathbf{p}_u + 2\mathbf{R}_u\mathbf{W_{opt}} = 0 \tag{4.20}$$

or equally

$$\mathbf{R}_u\mathbf{W_{opt}} = \mathbf{p}_u \tag{4.21}$$

Equation (4.21) represents the matrix form of the so-called Wiener-Hopf equation [34]. Pre-multiplying both sides of equation (4.21) by the inverse of the correlation matrix, i.e. $\mathbf{R_u}^{-1}$, we obtain the optimum weight vector

$$\mathbf{W_{opt}} = \mathbf{R_u}^{-1}\mathbf{p_u} \tag{4.22}$$

The optimum weight vector, $\mathbf{W_{opt}}$, is also called the *Wiener solution*. From equation (4.22), it is obvious that the computation of the optimum weight vector, $\mathbf{W_{opt}}$, requires knowledge of two quantities: the correlation matrix $\mathbf{R_u}$ of the input data vector $x(k)$, and the cross-correlation vector $\mathbf{p_u}$ between the input data vector $x(k)$ and the desired signal $d(k)$.

4.7.2 Method of Steepest-Descent

Although the Wiener-Hopf equation may be solved directly by calculating the product of the inverse of the correlation matrix $\mathbf{R}_u$ and the cross-correlation vector $\mathbf{p}_u$, this procedure results in serious computational difficulties since calculating the inverse of a matrix requires high computational complexity. An alternative procedure is to use the method of steepest-descent [34]. With this method the optimum weight vector $\mathbf{W_{opt}}$ is calculated in a recursive way, which means that the computations of the weight vectors proceed iteratively in a step-by-step manner. More specifically, the method begins with an initial value $\mathbf{W}(0)$ for the weight vector, which is chosen arbitrarily. Typically, $\mathbf{W}(0)$ is set equal to a column vector of an identity matrix. Using this initial or present guess, the gradient vector $\nabla(\mathbf{J}(w))$ is computed at time k, i.e., the kth iteration. The next guess of the weight vector is computed by making a change in the initial or present guess in a direction opposite to that of the gradient vector. The process is repeated until the algorithm converges onto the optimal value of the weight vector $\mathbf{W_{opt}}$, which would satisfy

$$\mathbf{J}(\mathbf{W_{opt}}) \leq \mathbf{J}(\mathbf{W}) \quad \text{for all } \mathbf{W} \tag{4.23}$$

Assuming there is a single minima, it is logical to consider that successive corrections to the weight vector in the direction of the negative of the gradient vector should eventually lead to the minimum mean-squared error $\mathbf{J}_{min}$, at which the weight vector assumes its optimum value $\mathbf{W}_{opt}$. Let $\mathbf{W}(k)$ denote the value of the weight vector at time k. According to the method of steepest-descent, the update value of the weight vector at time $k+1$ is computed by using the simple recursive relation

$$\mathbf{W}(k+1) = \mathbf{W}(k)\frac{1}{2}\mu[-\nabla(J(\mathbf{W}))] \tag{4.24}$$

where μ is a positive real-valued constant that is referred to as the *step-size parameter* or *weighting constant*. The factor $\frac{1}{2}$ is introduced for mathematical convenience.

From equation (4.20),

$$\nabla(J(\mathbf{W})) = -2\mathbf{p_u} + 2\mathbf{R_u}\mathbf{W}(k) \tag{4.25}$$

Substituting equation (4.24) into (4.23), we obtain

$$\mathbf{W}(k+1) = \mathbf{W}(k) + \mu[\mathbf{p_u} - \mathbf{R_u}\mathbf{W}(k)] \quad \text{for } k = 0, 1, 2 \ldots \tag{4.26}$$

The gradient vector in equation (4.24) may be also written in another form

$$\nabla(J(\mathbf{W})) = -2E[\mathbf{x}(k)d^*(k) - \mathbf{x}(k)\mathbf{x}^H(k)\mathbf{W}(k)] \qquad (4.27)$$
$$= -2E[\mathbf{x}(k)\{d(k) - y(k)\}^*] \qquad (4.28)$$
$$= -2E[\mathbf{x}(k)e^*(k)] \qquad (4.29)$$

Also, equation (4.23) can be expressed as

$$\mathbf{W}(k+1) = \mathbf{W}(k) + \mu E[\mathbf{x}(k)e^*(k)] \qquad (4.30)$$

It can be seen that the step-size parameter, μ, controls the size of the incremental correction applied to the weight vector as we proceed from one iteration cycle to the next. Hence, it determines the stability of the steepest descent method, together with the correlation matrix $\mathbf{R_u}$. In general, for stability and convergence of this algorithm

$$0 \leq \mu \leq \frac{2}{\lambda_{max}} \qquad (4.31)$$

where λ_{max} is the largest eigenvalue of $\mathbf{R}_u$.

Eigenvalues and eigenvectors result from solutions to the eigen problem, i.e., seeking non-trivial solutions to

$$\mathbf{A}\mathbf{x} = \lambda\mathbf{x} \qquad (4.32)$$

where scalar values of λ_n for which non-trivial solutions exist are referred to as the eigenvalues, and the corresponding solutions $\mathbf{x} \neq 0$ are the n eigenvectors (e_n). For example, λ_n and e_n for $\left(\begin{smallmatrix} 1 & a \\ a & 1 \end{smallmatrix}\right)$ are computed as follows. Solving the following equation gives the eigenvalues of the matrix $\mathbf{A}$.

$$det(\mathbf{A} - \lambda\mathbf{I}) \qquad (4.33)$$

Thus $\lambda_1 = 1 + a$ and $\lambda_2 = 1 - a$.

The eigenvectors corresponding to λ_1 are $\left(\begin{smallmatrix} x_1 \\ x_2 \end{smallmatrix}\right) = \left(\begin{smallmatrix} 1 \\ 1 \end{smallmatrix}\right)$ and the eigenvectors corresponding to λ_2 are $\left(\begin{smallmatrix} y_1 \\ y_2 \end{smallmatrix}\right) = \left(\begin{smallmatrix} 1 \\ -1 \end{smallmatrix}\right)$.

Equations (4.24) and (4.30) describe the mathematical formulation of the steepest-descent method. However, in a stationary environment the solution converges to the Wiener solution without having to invert the correlation matrix as explained in the following section where specific algorithms commonly employed to enable the method of steepest-decent are described.

4.7.3 Least-Mean-Squares Algorithm (LMS)

If it was possible to obtain exact measurements of the gradient vector $\nabla(\mathbf{J}(\mathbf{W}))$ at each iteration, and if the step-size parameter μ is suitably chosen, then the weight vector computed by using the steepest-descent method would indeed converge to the optimum Wiener solution. In reality, however, exact measurements of the gradient vector are not possible since this would require prior knowledge of both the correlation matrix $\mathbf{R_u}$ of the input data vector and the cross-correlation vector $\mathbf{p_u}$ between the input data vector and the desired signal. Consequently, the gradient vector must be estimated from the available data. In other words, the weight vector is updated in accordance with an algorithm that adapts to the incoming data. One such algorithm is the *least-mean-squares* (LMS) algorithm [34, 3]. A significant characteristic of the LMS algorithm is its simplicity. It does not require measurements of the relevant correlation functions, nor does it require matrix inversion.

To develop an estimate of the gradient vector $\nabla(\mathbf{J}(\mathbf{W}))$, the most obvious strategy is to substitute the expected value in equation (4.29) with the instantaneous estimate,

$$\nabla(\mathbf{J}(\mathbf{W})) = -2\mathbf{x}(k)e^*(k) \tag{4.34}$$

Substituting this instantaneous estimate of the gradient vector into equation (4.26) gives

$$\mathbf{W}(k+1) = \mathbf{W}(k) + \mu\mathbf{x}(k)e^*(k) \tag{4.35}$$

Now, the LMS algorithm can be presented by the following three equations

$$y(k) = \mathbf{W}^H(k)\mathbf{x}(k) \tag{4.36}$$

$$e(k) = d(k) - y(k) \tag{4.37}$$

$$\mathbf{W}(k+1) = \mathbf{W}(k) + \mu\mathbf{x}(k)e^*(k) \tag{4.38}$$

The LMS algorithm is a member of a family of stochastic gradient algorithms since the instantaneous estimate of the gradient vector is a random vector that depends on the input data vector $x(k)$. The LMS algorithm requires $2M + 1$ complex multiplications and $2M$ complex additions per iteration, where M is the number of weights (elements) used in the adaptive array. The response of the LMS algorithm is determined by three principal factors: the step-size parameter, the number of weights, and the eigen-value of the correlation matrix of the input data vector. A more detailed discussion about the LMS algorithm are available in the references [34], [48] and [49].

4.7.4   Direct Matrix Inversion (DMI) Algorithm

The *direct matrix inversion* (DMI) algorithm (also referred to as *sample matrix inversion* or SMI algorithm) estimates the array weights by replacing the correlation matrix $\mathbf{R_u}$ with its estimate. Using K samples $x(k)$, $k = 1, 2, \ldots, K-1$ of the array signals, an unbiased estimate of $\mathbf{R}$ may be obtained by means of a simple averaging scheme:

$$\hat{\mathbf{R}}_\mathbf{u}(k) = \frac{1}{K} \sum_{i=1}^{K-1} \mathbf{x}(k)\mathbf{x}^H(K) \qquad (4.39)$$

where $\hat{\mathbf{R}}_\mathbf{u}$ denotes the estimate at the kth instant of time and $x(k)$ denotes the array signal sample, also known as the *array snapshot*, at the kth instant of time (t is replaced by kT and the sampling time T is omitted for the ease of notation). The estimate of $\hat{\mathbf{R}}_\mathbf{u}$ may be updated when the new samples arrive using

$$\hat{\mathbf{R}}_\mathbf{u}(k+1) = \frac{k\hat{\mathbf{R}}_u(k) + \mathbf{x}(k+1)}{k+1} \qquad (4.40)$$

and a new estimate of the weights $W(k+1)$ at time instant $k+1$ may be made. The expression of the optimal weights requires the inverse of $\hat{\mathbf{R}}_\mathbf{u}$. This process of estimating $\mathbf{R_u}$ and then its inverse, i.e., $\hat{\mathbf{R}}_\mathbf{u}^{-1}$, may be combined to update the inverse of $\mathbf{R_u}$ from array signal samples using the *matrix inversion lemma* as follows:

$$\hat{\mathbf{R}}_\mathbf{u}^{-1}(k) = \mathbf{R}_\mathbf{u}^{-1}(k-1) - \frac{\mathbf{R}_\mathbf{u}^{-1}(k-1)\mathbf{x}(k)\mathbf{x}^H(K)\mathbf{R}_\mathbf{u}^{-1}(k-1)}{1 + \mathbf{x}^H(k)\mathbf{R}_\mathbf{u}^{-1}(k-1)\mathbf{x}(k)} \qquad (4.41)$$

with

$$\mathbf{R}_\mathbf{u}^{-1}(0) = \frac{1}{\varepsilon_0}\mathbf{I} \qquad (4.42)$$

This scheme of estimating weights using the inverse update is referred to as the *recursive least squares* RLS algorithm, which is further described in the next subsection. It should be noted that as the number of samples grows, the matrix update approaches its true value, and thus the estimated weights approach the optimal weights [33]. Hence, as

$$k \to \infty, \ \hat{\mathbf{R}}_\mathbf{u}^{-1}(k) \to \mathbf{R_u} \quad \text{and} \quad \mathbf{W}(k) \to \mathbf{W_{opt}} \qquad (4.43)$$

Discussion on the DMI algorithm may be found in [34] and [3]. Procedures for estimating array weights with efficient computation using DMI are considered in [50], and an analysis to show how it performs as a function of the number of snapshots is provided in [51]. Compared to the LMS algorithm, the convergence of the DMI algorithm is much faster. However, there is always a residual mean squared error in the DMI algorithm (due to estimation) that is greater than the LMS algorithm. Many applications of the DMI algorithm in mobile communications systems have been considered in numerous studies, for example

[52], [53], [54] and [55]. An application discussed in [54] is for vehicular mobile communications, while that presented in [55] is for reducing delay spread in indoor radio channels. A study referenced in [53] is for mobile satellite communications systems.

4.7.5 Recursive Least-Squares (RLS) Algorithm

Unlike the LMS algorithm which uses the method of steepest-descent to update the weight vector, the *recursive least squares* (RLS) algorithm [34, 48] uses the method of least squares to adjust the weight vector. In the method of least squares, we choose the weight vector $\mathbf{W}(k)$, so as to minimise a cost function that consists of the sum of squared errors over a time window. In the method of steepest-descent, on the other hand, we choose the weight vector to minimise the ensemble average of the squared errors. In the exponentially weighted RLS algorithm, at time k, the weight vector is chosen to minimise the cost function

$$\mathbf{J}(k) = \sum_{i=1}^{k} \lambda^{k-1} |e(i)|^2 \tag{4.44}$$

where $e(i)$ is defined in equation (4.9), and λ is a positive constant close to, but less than, one, which determines how quickly the previous data are de-emphasised. In a stationary environment, however, λ should be equal to 1, since all data past and present should have equal weight. The RLS algorithm is obtained from minimising equation (4.44) by expanding the magnitude squared and applying the matrix inversion lemma. The RLS algorithm can be described by the following equations [34]

$$\mathbf{K}(k) = \frac{\lambda^{-1}\mathbf{P}(k-1)\mathbf{x}(k)}{1 + \lambda^{-1}\mathbf{x}^H(k)\mathbf{P}(k-1)\mathbf{x}(k)} \tag{4.45}$$

$$\xi(k) = d(k) - \mathbf{W}^H(k-1)\mathbf{x}(k) \tag{4.46}$$

$$\mathbf{W}(k) = \mathbf{W}(k-1) + \mathbf{K}(k)\xi(k) \tag{4.47}$$

$$\mathbf{P}(k) = \lambda^{-1}\mathbf{P}(k-1) - \lambda^{-1}\mathbf{K}(k)\mathbf{x}^H(k)\mathbf{P}(k-1) \tag{4.48}$$

The initial value of $\mathbf{P}(k)$ can be set to

$$\mathbf{P}(0) = \delta^{-1}\mathbf{I} \tag{4.49}$$

where $\mathbf{I}$ is the $M \times M$ identity matrix, and δ is a small positive constant called the *regularisation parameter*, which is assigned with a small value for high SNR and a large value for low SNR. An important feature of the RLS algorithm is that it utilises information contained in the input data, extending back to the instant of time when the algorithm is initiated. The resulting rate of convergence is therefore typically an order of magnitude faster than the simple LMS algorithm. This improvement in performance, however, is achieved at the expense of a large increase in computational complexity. The RLS algorithm requires $4M^2 + 4M + 2$ complex multiplications per iteration, whereas the LMS algorithm requires 2M+1 complex multiplications and 2M complex additions per iteration, where M is the number of weights used in the adaptive array.

4.8 SOURCE LOCATION TECHNIQUES

As well as for spatial reference beamforming, signal direction finding based on *direction-of-arrival* (DoA) estimation using an antenna array is considered another major application for antenna array processing since it enables users to be located and this information can to useful to both the user, service provider and authorities (such an example is given in chapter 6). DoA estimation techniques can be classified by three main types: conventional, subspace and maximum likelihood.

The classical beamformer structure, as shown in figure 4.7, has an output signal $y(k)$ (4.50) and is given by a linearly weighted sum of the sensor element.

$$y(k) = \mathbf{W}^H \mathbf{U}(k) \tag{4.50}$$

where $\mathbf{W}$ is the array weight vector and $\mathbf{U}$ is the received signal vector. The total output of the conventional beamforming is given by the following formula

$$P_{cbf} = E\left[|y(k)|^2\right] \tag{4.51}$$

$$= E\left[\left|\mathbf{W}^H \mathbf{U}(k)\right|^2\right] \tag{4.52}$$

$$= \mathbf{W}^H E\left[\mathbf{U}(k)\mathbf{U}^H(k)\right]\mathbf{W} \tag{4.53}$$

$$= \mathbf{W}^H \mathbf{R_u} \mathbf{W} \tag{4.54}$$

where $\mathbf{R_u}$ is the spatial correlation matrix $\mathbf{R} = E\left[\mathbf{U}(t)\mathbf{U}^H(t)\right]$ of the array input data. The above equation is the of basis for all the conventional DoA algorithms. Conventional methods are based on using beamforming and null-steering to scan through the spatial power spectrum to identify power peaks that correspond to valid signal direction of arrivals, i.e., a beam is scanned across the array and the output signal power, $|y(\theta)|^2$ is plotted as a function of angle. The Fourier and Capon algorithms are classified here as conventional. Subspace-based methods for DoA estimation were proposed in order to solve the main disadvantages of classical beamforming based methods that are limited in angular resolution, by exploiting the structure of the input array data model. The two main subspace-based techniques are called the *Multiple SIgnal Classification* (MUSIC) algorithm and the *Estimation of Signal Parameters via Rotational Invariance Technique* (ESPRIT). *Maximum likelihood* (ML) techniques achieve better performance than subspace based techniques, especially in low signal-to-noise ratio conditions or when only a small number of signal samples are available. In addition, ML based techniques can also perform well in correlated signal conditions but are computationally intensive compared to conventional and subspace based techniques.

Several DoA estimation techniques are compared in terms of spatial resolution later in this chapter.

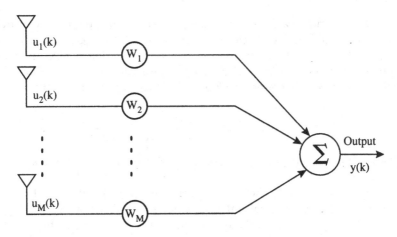

Fig. 4.7 Classical beamforming structure.

4.9 FOURIER METHOD

In the *Fourier method* (FM) [33], also referred to as the *classical beamformer*, *periodogram* or the *delay-and-sum* method, the beam is scanned over the angular region of interest in discrete steps by forming weights $W = \alpha(\phi)$. Weight $W = \alpha(\phi)$ is the steering vector associated with the DoA angle ϕ, and for different angles the output power is measured. So, the output power at the classical beamformer as a function of the *angle of arrival* (AoA) is given by

$$P_{cbf}(\phi) = \mathbf{W}^H \mathbf{R_u} \mathbf{W} = \alpha^H(\phi)\mathbf{R_u}\alpha(\phi) \tag{4.55}$$

The angle of arrival can be estimated by finding the angles that correspond to the peaks in the output power. Hence, by estimating the input autocorrelation matrix $\mathbf{R_u}$ and knowing the steering vectors $\alpha(\phi)$ for all ϕ's, we can estimate the output power as function of the AoA. The Fourier method has many disadvantages. In cases where signals arrive from multiple directions and sources, the width of the beam and the size of the sidelobes limit the effectiveness leading this technique into poor resolution results. The method performs well under the presence of a single signal. When there is more than one signal present, the array output not only contains contributions from the desired signal, but also from the undesired ones from other directions as well.

4.10 CAPON'S MINIMUM VARIANCE

Capon's minimum variance technique [56] makes an attempt to improve the poor resolution associated with the Fourier method by using some of the degrees of freedom to form a beam in the desired look direction whilst at the same time using the remaining degrees of freedom to form nulls in the direction of interfering signals. The output power for Capon's method as a function of the AoA is given by

$$P(\phi) = \frac{1}{\alpha^H(\phi) \mathbf{R_u}^{-1} \alpha(\phi)} \tag{4.56}$$

The angles of arrival can be estimated by locating peaks in the spatial spectrum $P(\phi)$.

Despite providing a better resolution compared with the Fourier method, Capon's technique has several disadvantages. When other signals present are correlated with the signal of interest, the correlated components may be combined destructively and Capon's method fails. Finally, estimation of DoA using Capon's method requires the computation of a matrix inverse which can increase the computational cost for large arrays.

4.11 THE MUSIC ALGORITHM

The MUSIC algorithm [57] is based on exploiting the eigenstructure of the input spatial covariance matrix and provides high resolution. The MUSIC algorithm requires very precise and accurate array calibration and gives information about the number of incident signals, DOA of each signal, strengths and cross-correlations between the incident signals and noise power.

If there are D signals incident on the array, the received input data vector at an M-element array can be expressed as a linear combination of the D incident waveforms and noise. So,

$$U(t) = \sum_{l=0}^{D-1} \alpha(\phi_l) s_l(t) + n(t) \tag{4.57}$$

$$U(t) = [\alpha(\phi_0)\,\alpha(\phi_1)\dots\alpha(\phi_{D-1})] \begin{bmatrix} s_0(t) \\ \dots \\ s_{D-1}(t) \end{bmatrix} + n(t) = \mathbf{A}s(t) + \mathbf{n}(t) \quad (4.58)$$

Where $\mathbf{s}^T(t) = [s_0(t)\,s_1(t)\dots s_{D-1}(t)]$ is the vector of incident signals, $\mathbf{n}(t) = [\,n_0(t)\quad n_1(t)\quad\dots\quad n_{D-1}(t)\,]$ is the noise vector and $\alpha(\phi_j)$ is the array steering vector which corresponds to the DoA of the jth signal. The input covariance matrix can be expressed as $\mathbf{R_u}$ and is $\mathbf{R_u} = E\left[\mathbf{UU}^H\right]$ estimated by

$$\hat{\mathbf{R}}_\mathbf{u} = \frac{1}{k}\sum_{k=0}^{K-1}\mathbf{U}_k\mathbf{U}_k^H \quad (4.59)$$

where $\mathbf{U}_k$, $k = 0,1,\dots,k-1$ are the input samples (as described at the beginning of this chapter). Then, eigen decomposition is performed on the estimate of $\mathbf{R_u}$

$$\hat{\mathbf{R}}_\mathbf{u}\mathbf{V} = \mathbf{V}\mathbf{\Lambda} \quad (4.60)$$

where $\mathbf{\Lambda} = diag\{\ \lambda_0\quad \lambda_1\quad\dots\quad\lambda_{M-1}\ \}$ are the eigenvalues and $\lambda_0 \geq \lambda_1 \geq \dots \geq \lambda_{M-1}$ are the corresponding eigenvectors of

$$\mathbf{V} = [\ q_0\quad q_1\quad\dots\quad q_{M-1}\ ] \quad (4.61)$$

The number of signals D, from the multiplicity K, of the smallest eigenvalue λ_{min} is estimated as

$$\hat{D} = M - K \quad (4.62)$$

the MUSIC spectrum is computed

$$\hat{P}_{MUSIC}(\phi) = \frac{\alpha^H(\phi)\,\alpha(\phi)}{\alpha^H(\phi)\,\mathbf{V}_n\mathbf{V}_n^H\alpha(\phi)} \quad (4.63)$$

where $\mathbf{V}_n = [\ q_D\quad q_{D+1}\quad\dots\quad q_{M-1}\ ]$ and we find the D largest peaks of $\hat{P}_{MUSIC}(\phi)$ in order to estimate the DoA.

The MUSIC algorithm is based on the observation that one can estimate the steering vectors associated with the received signals by finding the steering vectors which are most nearly orthogonal to the eigenvectors associated with the eigenvalues of $\mathbf{R_u}$.

In contrast to conventional methods, the MUSIC spatial spectrum does not estimate the signal power associated with each arrival angle. Under uncorrelated and identical noise conditions, the peaks of $P_{MUSIC}(\phi)$ are guaranteed to correspond to the true DoA. When array calibration is perfect, even closely spaced signals can be distinguished and resolved. However, MUSIC fails when impinging signals $s(t)$ are highly correlated.

Many improvements and modifications to the MUSIC algorithms have been proposed to increase the resolution and decrease the computational complexity.

One of these improvements is the Root-MUSIC algorithm [58]. Root-MUSIC is based on polynomial rooting. Although it can only be applied to a uniformly spaced linear array, it provides greatly improved resolution over conventional MUSIC, especially in low SNR conditions. As defined previously, the MUSIC spatial spectrum is expressed as

$$P_{MUSIC}(\phi) = \frac{1}{\alpha^H(\phi)\mathbf{V}_n\mathbf{V}_n^H\alpha(\phi)} \tag{4.64}$$

or

$$= \frac{1}{\alpha^H(\phi)\mathbf{Q}\alpha(\phi)} \tag{4.65}$$

Where $\mathbf{Q} = \mathbf{V}_n\mathbf{V}_n^H$. For a uniformly spaced linear array, the mth element of the steering vector $\alpha(\phi)$ can be defined as

$$\alpha_m(\phi) = e^{(j\beta md \cos \phi)} \tag{4.66}$$

Where d is the spacing between the consecutive antenna elements and $\beta = \frac{2\pi}{\lambda}$ phase propagation factor. Making use of this, the reciprocal of the MUSIC spatial spectrum can be rewritten as

$$P_{MUSIC}^{-1}(\phi) = \sum_{n=1}^{M} \sum_{m=1}^{M} e^{(-j\beta md \cos \phi)}\mathbf{Q}_{nm}e^{(j\beta md \cos \phi)} \tag{4.67}$$

Or simplified to

$$P_{MUSIC}^{-1}(\phi) = \sum_{p=-M+1}^{M-1} \mathbf{Q}_p e^{(-j\beta p \Delta x \cos \phi)} \tag{4.68}$$

When the two summations are combined, where $\mathbf{Q}_p = \sum_{m-n=p}\mathbf{Q}_{nm}$, i.e., the pth diagonal entries of $\mathbf{Q}$.

By defining a polynomial $P(z)$

$$P(z) = \sum_{p=-M+1}^{M-1} Q_p z^{-1} \tag{4.69}$$

The MUSIC spectrum can be evaluated by resolving the $P(z)$ on a unit circle, the peaks of the spectrum equate to the roots of the $P(z)$ close to the unit circle, i.e. the kth pole (corresponding to the kth signal) of $P(z)$ at $z = z_k = |z_k|\exp(j \arg z_k)$ leading to

$$\phi_k = \cos^{-1}\left(\frac{\arg(z_k)}{\beta d}\right) \tag{4.70}$$

Further types of MUSIC variation can be found in the literature. Many are based on further exploiting the signal properties for better accuracy [59], others are based on improving the speed and efficiency of the subspace decomposition [60]. The selection of an appropriate algorithm is therefore highly dependant on the requirements and conditions of deployment.

4.12 ESPRIT

The ESPRIT algorithm [44] is a computationally efficient and robust method for DoA estimation and does not require an exhaustive search through all possible steering vectors to estimate the DoA of the incoming signal. Here, the M elements of the receiving array are divided into two identical overlapping sub-arrays, each of which consists of the $M-1$ element sensor doublets (pairs) displaced by a known constant displacement vector Δx that sets the reference direction, and all angles are measured with reference to this vector, e.g., for a four-element ULA the array is divided into two sub-arrays of three elements. The sub-arrays consist of two doublets spaced apart by d, as shown in figure 4.8. The output of each sub-array is denoted by $U_0(t)$ and $Uu_1(t)$. Using matrix and vector notation these two outputs can be written as

$$U_0(t) = \mathbf{A}\mathbf{s}(t) + \mathbf{n}(t) \tag{4.71}$$

$$U_1(t) = \mathbf{A}\mathbf{\Phi}\mathbf{s}(t) + \mathbf{n}_1(t) \tag{4.72}$$

where $\mathbf{s}(t)$ denotes the D source signals observed at a reference element, $\mathbf{n}_0(t)$ and $\mathbf{n}_1(t)$ denote the noise present on the elements of the two sub-arrays and $\mathbf{A}$ denotes a $(M-1) \times D$ matrix, with its columns denoting the D steering vectors corresponding to D directional sources associated with the first sub-array. The steering vectors corresponding to D directional sources associated with the second sub-array are given by $\mathbf{A}\mathbf{\Phi}$ where $\mathbf{\Phi}$ is an $D \times D$ diagonal matrix whose diagonal elements represent the phase delays between the doublet sensors for the D signals.

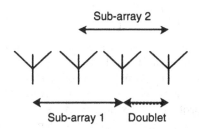

Fig. 4.8 Sub-array arrangement.

Let $\mathbf{Ux}$ and $\mathbf{Uy}$ denote two $(M-1) \times D$ matrices with their columns denoting the D eigenvectors corresponding to the largest eigenvalues of the two array covariance matrices $\mathbf{Rxx}$ and $\mathbf{Ryy}$, respectively. As these two sets of eigenvectors span the same D-dimensional signal space, it follows that these two matrixes $\mathbf{Ux}$ and $\mathbf{Uy}$ are related by a unique non-singular transformation matrix $\mathbf{\Psi}$, thus

$$\mathbf{Uy} = \mathbf{Ux}\mathbf{\Psi} \tag{4.73}$$

Similarly, these matrices are related to steering vector matrixes A and $A\Phi$ by a unique non-singular transformation matrix, which is

$$\mathbf{Ux} = \mathbf{AT} \tag{4.74}$$

$$\mathbf{Uy} = \mathbf{A\Phi T} \tag{4.75}$$

Substituting for $\mathbf{Ux}$ and $\mathbf{Uy}$, and the fact that $\mathbf{A}$ is of full rank yields

$$\mathbf{T\Psi T}^{-1} = \mathbf{\Phi} \tag{4.76}$$

Which states that the eigenvalues of Ψ are equal to the diagonal elements of Φ and that the columns of $\mathbf{T}$ are eigenvectors of Ψ. This is the main relationship in the development of ESPRIT. An eigen-decomposition of Ψ gives its eigenvalues, and by equating them to Φ leads to the DoA estimation, so

$$\hat{\phi}_k = \cos^{-1}\left[\frac{(\arg \hat{\Phi}_k)}{\beta d}\right] \tag{4.77}$$

The technique described above is based on the rotational invariance principle, i.e., a linear phase shift along the array is assumed. As previously mentioned, the operation of ESPRIT is based upon the array having a structure that can be decomposed into a number of identical sub-arrays with corresponding elements of each sub-array displaced by a linear distance dx. The sensors should occur in matched pairs with identical displacement. Hence, this algorithm is well suited to uniform linear arrays.

4.12.1 Unitary ESPRIT

Unitary ESPRIT is a variation of standard ESPRIT that has been summarised in the previous section and is shown in [61] to obtain increased estimation accuracy and reduced computational complexity compared to standard ESPRIT. The reduced complexity is achieved by transforming the subspace estimation of $\mathbf{R_u}$ into a real-valued problem by exploiting the structure of centro-symmetric arrays (such as a uniform linear array where the element locations are symmetric with respect to the centroid). This property constrains the rotational invariance matrix (Φ) to the unit circle which can also increase estimation accuracy as it reduces estimation errors. The algorithm also contains a DoA reliability test without additional complexity. The algorithm can be summarised as follows [62]. Given a centro-symmetric array structure, the array steering vector, $\mathbf{A}$, satisfies $\Pi_{\mathbf{M}}\mathbf{A}=\bar{A}\Lambda$ for some unitary diagonal matrix, Λ, where $\Pi_{\mathbf{M}}$ is an $M \times M$ exchange matrix with ones along the anti-diagonal and zeros elsewhere and $\bar{A}$ denotes complex conjugate without transposition. In the first step of unitary ESPRIT, forward-backward averaging is incorporated by transforming the complex valued received signal matrix, $\mathbf{X}$, into a real valued matrix, $\mathbf{T(X)}$ (denoted in matrix form as $\mathbf{T}$)

given by equation (4.78) where $\mathbf{Q}$ is a left-hand real matrix satisfying $\Pi\bar{Q}=\mathbf{Q}$ and is fully defined in [62]. Note that forward-backward averaging helps to de-correlate coherent multipath signals that can otherwise reduce dimensionality of the signal subspace and cause the algorithm to fail. Standard ESPRIT also allows forward-backward averaging to be incorporated by computing the spatial covariance matrix from a number of sub-arrays.

$$T(X) = \mathbf{Q_M^H}[\mathbf{X\Pi_M\bar{X}\Pi_M}]\mathbf{Q_{2N}} \tag{4.78}$$

Now, for a large number of snapshots, the DoA (ϕ) can be determined from equation (4.79) for D dominant eigenvectors of $\mathbf{T}$. D denotes the model order that is estimated during the early stages of the algorithm and is computed from the spatial correlation matrix using an iterative technique similar to that described in reference [63]. The model order is closely related to the number of signals impinging on the array. The technique is based on the idea that the smallest eigenvector of the spatial correlation matrix is larger than the largest eigenvector of the remaining terms. Thus, eigenvectors are selected in descending order.

$$\Omega = diag\{\lambda_k\}_{k=1}^{D} = diag\left\{ \tan \frac{\phi_k}{2} \right\}_{k=1}^{D} \tag{4.79}$$

Where Ω is obtained from the eigen-decomposition of equation (4.80) and e_k is the kth eigenvalue and $\phi_k = 2\tan_{-1}\omega_k$.

$$\mathbf{Y} = \mathbf{T\Omega T}^{-1} \tag{4.80}$$

The reliability test is incorporated into unitary ESPRIT and the DoA estimate can be considered as reliable if the eigenvalues of $\mathbf{Y}$ are real. Alternatively, if they occur in complex-conjugate pairs, the test has failed and the algorithm should be re-run using an increased number of signal estimates.

Finally, equation (4.81) computes the AoA of the D impinging signals where d is the inter-element spacing and e_d is the dth eigenvalue of equation (4.80). However, a power azimuth spectrum also requires the received signal powers associated with each angle. This is achieved by constructing a matrix of beamformer weight vectors, $\mathbf{W}$, which are orthogonal with respect to the estimated directions of arrival, i.e., they minimise the power at the output of the beamformer. The path powers are then defined as the pseudo-inverse of $\mathbf{W}$. This, in theory, leads to an optimum solution but may be perturbed by additional measurement noise.

$$\hat{\phi}_d = \sin^{-1}\left(\frac{\lambda}{\pi d} \tan^{-1}(e_d)\right) \tag{4.81}$$

4.13 MAXIMUM LIKELIHOOD TECHNIQUES

In geometric terms, the *maximum likelihood* (ML) estimator is obtained by searching over the array manifold for the D steering vectors that form a D-dimensional signal subspace which is closest to the array input data vectors, where the closeness is measured by the modulus of the projection of the vectors onto this subspace.

The basic ML estimator is based on maximisation of the following log-likelihood function with respect to the unknown parameters

$$J = -ND\log\sigma^2 - \frac{1}{\sigma^2} \sum_{t=0}^{N-1} |\mathbf{U}(t) - \mathbf{A}(\phi)\mathbf{s}(t)|^2 \qquad (4.82)$$

Where $\mathbf{A}(\phi) = [a\,(\phi_0)\,a\,(\phi_1)\,......a\,(\phi_{D-1})]$ is the spatial signature vector, $\mathbf{U}(t)$ is the received signal vector, $\mathbf{s}(t)$ denotes the D source signals observed, N is the number of snapshots considered and σ^2 is the noise covariance. This can be represented equivalently as following the minimisation function form

$$\min_{(\phi,s)} \left\{ \sum_{t=0}^{N-1} |\mathbf{U}(t) - \mathbf{A}(\phi)\mathbf{s}(t)|^2 \right\} \qquad (4.83)$$

Fixing ϕ and minimising with respect to $\mathbf{s}$ leads to the least square solution

$$\hat{\mathbf{s}}(t) = (\mathbf{A}^H(\phi)\mathbf{A}(\phi))^{-1}\mathbf{A}^H(\phi)\mathbf{U}(t) \qquad (4.84)$$

Substituting (4.84) into (4.83) gives

$$\min_{(\phi)} \sum_{t=0}^{N-1} |\mathbf{U}(t) - \mathbf{P}\mathbf{u}(t)|^2 \qquad (4.85)$$

where $\mathbf{P}$ is the projection operator which projects onto the space spatial signature matrix $\mathbf{A}(\phi)$ given by

$$\mathbf{P} = \mathbf{A}(\phi)(\mathbf{A}^H(\phi)\mathbf{A}(\phi))^{-1}\mathbf{A}^H(\phi) \qquad (4.86)$$

The maximum likelihood (ML) estimate for the DoAs can therefore be obtained by maximising the log-likelihood function

$$J(\phi) = \sum_{t=0}^{N-1} |\mathbf{P}\mathbf{U}(t)|^2 \qquad (4.87)$$

Using geometric terms it can be shown that the log-likelihood function can be expressed as follows

$$J(\phi) = trace\,[\mathbf{P}\mathbf{R}_u] \qquad (4.88)$$

Where $\mathbf{R}_u$ is the spatially sampled covariance matrix.

The maximisation of a log-likelihood function is a nonlinear multi-dimensional problem and therefore computationally very intense. Many modifications to the ML estimator have been proposed with the goal of reducing the computational load and simplifying the maximisation solution. The *alternating projection algorithm* [64], as an example, uses an iterative relaxation based technique by maximising the equation using a single parameter whilst fixing all other parameters. This reduces the multi-dimensional maximisation problem to a one-dimensional problem. Other types of ML estimator can be found in the literature and one commonly referenced ·algorithm is the *space alternating generalised expectation maximisation* (SAGE) algorithm [65].

4.14 SPATIAL SMOOTHING

In multipath signal conditions, signals arriving at the base station can have a high degree of correlation. In these conditions the covariance matrix $\mathbf{R}_u$ approaches a singular state, thereby preventing subspace based DoA estimation techniques from functioning. To circumvent this problem suitable preprocessing termed *spatial smoothing* [66] can be used to de-correlate the signals before a technique like MUSIC or ESPRIT is used. Spatial smoothing can be done by averaging the covariance matrix $\mathbf{R}_u$ over overlapping arrays (similar to that used for the unitary ESPRIT algorithm). A linear uniform array with M sensors is divided into L overlapping forward sub-arrays of size p, in a way that the sensor elements $0, \ldots, p-1$ form the first forward sub-array while sensors $1, \ldots, p$ form the second forward sub-array, etc. Denote by $\mathbf{U}_k(t)$ the vector of the received signals at the kth forward sub-array and the signals received at each sub-array are

$$\mathbf{U}_k^f(t) = \mathbf{A}\mathbf{D}^{(k-1)}\mathbf{s}(t) + \mathbf{n}_k(t) \tag{4.89}$$

where $\mathbf{D}^{(k)}$ denotes the kth power of the diagonal matrix

$$\mathbf{D} = \begin{bmatrix} \exp\left(-j\beta\cos\left(\varphi_0\right)\right) & \ldots & 0 \\ \ldots & \ldots & \ldots \\ 0 & \ldots & \exp\left(-j\beta\cos\left(\varphi_{L-1}\right)\right) \end{bmatrix} \tag{4.90}$$

The covariance matrix of the kth forward sub-array is given by

$$\mathbf{R}_k^f = \mathbf{A}\mathbf{D}^{(k-1)}\mathbf{R}_{ss}\mathbf{D}^{H(k-1)}\mathbf{A}^H + \sigma_n^2\mathbf{I} \tag{4.91}$$

where $\mathbf{R}_{ss}$ is the covariance matrix of the sources. Therefore the forward averaged spatially smoothed covariance matrix $\mathbf{R}^f$ can be defined as the average of the sub-array covariance matrixes

$$\mathbf{R}^f = \frac{1}{L}\sum_{k=0}^{L-1}\mathbf{R}_k^f \tag{4.92}$$

where $L = M - p + 1$ is the number of subarrays. Consequently, from (4.91) and (4.92) we get

$$\mathbf{R}^f = \mathbf{A}\left(\frac{1}{L}\sum_{k=0}^{L-1}\mathbf{D}^{(k-1)}\mathbf{R}_{ss}^f\left(\mathbf{D}^{(k-1)}\right)^H\right)\mathbf{A}^H + \sigma_n^2\mathbf{I} \qquad (4.93)$$

where $\mathbf{R}_{ss}^f$ represents the modified covariance matrix of the signals equal to

$$\mathbf{R}_{ss}^f = \frac{1}{L}\sum_{k=0}^{L-1}\mathbf{D}^{(k-1)}\mathbf{R}_{ss}\left(\mathbf{D}^{(k-1)}\right)^H \qquad (4.94)$$

If the number of sub-arrays is bigger than the number of incident signals the modified covariance matrix of the signals will be non-singular regardless of the coherence of the signals. The use of spatial smoothing reduces the effective size of the array, so results in a reduced array aperture. Forward-averaging spatial smoothing before the MUSIC algorithm makes it possible to detect $M/2$ high correlated signals in contrast with the $M - 1$ low correlated signals that can be detected by conventional MUSIC. Now let overlapping backward sub-arrays of p elements such that the first backward sub-array is formed using elements $M, M - 1, \ldots, M - p + 1$, the second $M - 1, M - 2, \ldots, M - p$ elements, etc. The complex conjugate of the received signal vector at the kth backward sub-array is given by

$$\mathbf{U}_k^b = \left[\mathbf{U}_{M-k+1}^*, \mathbf{U}_{M-k}^*, \cdots, \mathbf{U}_{p-k+1}^*\right]^T \qquad (4.95)$$

$$= \mathbf{A}\mathbf{D}^{k-1}\left(\mathbf{D}^{M-1}\mathbf{s}\right)^* + \mathbf{n}_k^*, \qquad (4.96)$$

$$0 \leq k \leq L - 1 \qquad (4.97)$$

where D is defined in (4.90). The covariance matrix of the kth backward sub-array is therefore given by

$$\mathbf{R}_k^b = \mathbf{A}\mathbf{D}^{k-1}\mathbf{R}_{bss}\left(\mathbf{D}^{k-1}\right)^H\mathbf{A}^H + \sigma_n^2\mathbf{I} \qquad (4.98)$$

where

$$\mathbf{R}_{bss} = \mathbf{D}^{-(M-1)}E\left[\mathbf{s}^*\mathbf{s}^T\right]\left(\mathbf{D}^{-(M-1)}\right)^H \qquad (4.99)$$

$$= \mathbf{D}^{-(M-1)}\mathbf{R}_{bss}^*\left(\mathbf{D}^{-(M-1)}\right)^H \qquad (4.100)$$

The backward conjugate averaged spatially smoothed covariance matrix $\mathbf{R}^b$ will be:

$$\mathbf{R}^b = \frac{1}{L}\sum_{k=0}^{L-1}\mathbf{R}_k^b = \mathbf{A}\mathbf{R}_{ss}^b\mathbf{A}^H + \sigma_n^2\mathbf{I} \qquad (4.101)$$

The backward conjugate averaged spatially smoothed covariance matrix $\mathbf{R}^b$ has the same properties as the forward one and the forward/conjugate backward smoothed covariance matrix is defined as

$$\mathbf{R}^{f/b} = \frac{\mathbf{R}^f + \mathbf{R}^b}{2} \tag{4.102}$$

Applying MUSIC on $\mathbf{R}^{f/b}$, it is possible to detect up to $2M/3$ coherent signals.

4.14.1 Comparison of Spatial Parameter Estimation Techniques

Figure 4.9 shows the spatial spectrum generated by the MUSIC, ESPRIT, Capon and delay-sum techniques for equal amplitude signals arriving at $-7°$, $0°$ and $23°$ and for an eight-element array. The amplitudes of the spatial spectrums have been normalised for convenient comparison. The figure demonstrates the achievable resolution from each technique, where the Capon and delay-sum method is unable to resolve the signals at $-7°$ and $0°$, although the Capon method does show some distinction between the signals. Conversely, the super-resolution techniques are able to distinguish between the signals. The output of ESPRIT is also shown to produce the discontinuous spectrum with impulses appearing at the desired angles. Finally, the MUSIC algorithm is shown to produce a continuous spectrum with distinctive peaks at the angles of the arriving signals.

4.15 DETERMINATION OF NUMBER OF SIGNAL SOURCES

Most DoA estimation algorithms require knowledge of the number of signals impinging on the antenna array. For subspace-based techniques especially, the estimation of the number of signal sources can be considered a key step, where error in the estimation process will lead to a sub-optimum DoA estimation. A simple detection can be done by estimating the multiplicity, K, of the smallest eigenvalue, $\lambda_{\min}$, of the covariance matrix $\mathbf{R}_u$, and correspond to the non-principal eigen subspace (noise subspace). The value of these K eigenvalues should be equal to the noise variance σ_n^2. The estimated number, L, of the incident signals is then given by $L = M - K$. However, in practice, all the eigenvalues corresponding to the noise power are not identical since the covariance matrix $\mathbf{R}_u$ is estimated from a finite number of data samples. Using the *Akaika information theoretic criteria* (AIC) [67], the number of sources is determined by the argument that minimises the follow criterion

$$AIC(d) = -\log \left[\frac{\prod\limits_{i=d+1}^{M} \lambda_i^{\frac{1}{(M-d)}}}{\frac{1}{M-d} \sum\limits_{i=d+1}^{M} \lambda_i} \right]^{(M-d)N} + d(2M - d) \tag{4.103}$$

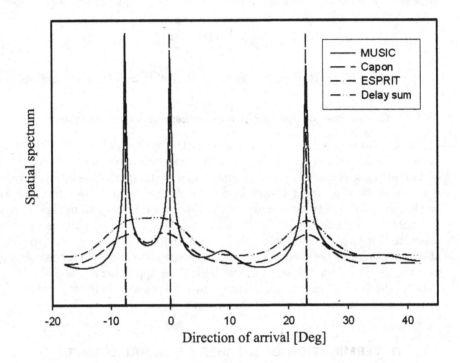

Fig. 4.9 Spatial spectrum comparison.

Where λ_i are the eigenvalues of the input spatial covariance matrix $\mathbf{R}_u$, M is the number of elements and N is the number of snapshots. The first term in equation (4.103) is derived from the log-likelihood function by taking the closeness of the eigenvalues as a measure of the ratio of their geometric mean to their arithmetic mean. The second term of the equation is a penalty factor added by the AIC criterion.

Alternatively, using the minimum descriptive length criteria approach [68], the number of sources is determined as the argument that minimises the follow criterion

$$MDL(d) = -\log \left[\frac{\prod_{i=d+1}^{M} \lambda_i^{\frac{1}{(M-d)}}}{\frac{1}{M-d} \sum_{i=d+1}^{M} \lambda_i} \right]^{(M-d)N} + \frac{1}{2}d(2M-d)\log N \qquad (4.104)$$

As with the AIC equation, the first term of the equation here is derived from the log-likelihood function, while the second term is a penalty factor added by the MDL criterion.

If spatial smoothing is applied, the AIC and MDL criterion have to be modified since the spatial smoothing processing complicates signal order estimation. The penalty factors are modified as follows. For forward averaging, the AIC penalty factor becomes $d(2M - 2d + 1)$ and the MDL penalty factor becomes $0.5d(2M - d + 1)\log N$. For forward/conjugate backward smoothing the AIC penalty factor is updated to $0.5d(2M - 2d + 1)$ and the MDL penalty factor to $0.25d(2M - d + 1)\log N$.

4.16   BLIND BEAMFORMING

The class of blind adaptive algorithms comprises such algorithms that do not require a training signal or information on the geometrical properties of the array. Instead, they exploit some known structural and statistical properties of the transmitted signal or of the desired received signal, such as the *constant modulus* property or the *spectral self-coherence* property. Since the non-blind algorithms use a training signal, during the training period, data cannot be sent over the radio channel; hence the spectral efficiency of the system is reduced. Therefore, the blind algorithms have an important role in upgrading the performance of a system in terms of link efficiency.

In the DoA estimation-based beamforming algorithms, the DoAs of the received signals are first determined by using prior knowledge of the array response. After the DoAs are estimated, an optimum beamformer is then constructed from the corresponding array response to extract the desired signals from interference and noise. The performance of this technique strongly depends on the reliability of the prior spatial information, e.g., the array manifold. In many situations of practical interest, this information is not available due to inaccurate calibration or coherent multipath signals. Even if this information is available, the cost and complexity can be high and the information may be inaccurate. The computational complexity of these algorithms is also very high. In a wireless communication system, especially in a code division multiple access (CDMA) system, the number of users of a radio channel may be greater than the number of array elements, and if the multipath of each user's signal is also taken into account, the total number of signals impinging on the array will far exceed the number of array elements (especially since it is impractical to employ large arrays in these scenarios). In this case the DoA estimation algorithm will fail. Thus direct application of DoA estimation to control a beamformer is unlikely to be used for the reception of uplink signals.

In the case of a receiving system using a blind beamforming algorithm that exploits neither a training sequence nor the properties of the receiver array, the algorithm has as its input the received signals at the antenna elements sampled in the time domain, i.e. the array output data matrix $\mathbf{X}$. From this, the blind algorithm tries to extract the unknown channel impulse response $\mathbf{H}$ and the unknown transmitted data $\mathbf{S}$. This is illustrated in the simple block diagram in figure 4.10.

In principle, there are many possible decompositions of $\mathbf{X}$. The algorithm uses additional prior knowledge of the characteristics. Even though it does not know the actual 'bits', the receiver knows that, for example, the transmitted signal must have a constant envelope, and thus tries to find that decomposition of $\mathbf{HS}$ where $\mathbf{S}$ really exhibits a constant envelope. Also other statistical properties of the modulated signal, like the *finite alphabet of symbols*, or even *cyclostationarity*, can be used. Currently, all blind algorithms require too many computations to be applied in real-time systems. Thus, the use of training sequences in so-called semi-blind algorithms speeds up computation considerably compared to blind techniques. A recent algorithm of this class, DILFAST (*Decoupled Iterative Least Squares Finite Alphabet Space Time*) has real-time capability [1] and is briefly described in the following section.

Note that the goal of blind beamforming algorithms is the same as all the algorithms described so far, i.e., to enhance the received signal by means of adjusting the complex array weights.

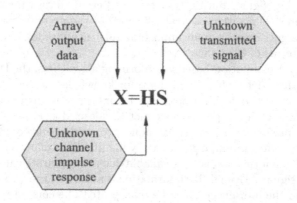

Fig. 4.10 Basic concept of blind beamforming algorithms.

4.16.1 Decoupled Iterative Least Squares Finite Alphabet Space-Time (DILFAST) Algorithm

The DILFAST (*Decoupled Iterative Least Squares Finite Alphabet Space-Time*) algorithm is a semi-blind detection scheme which adaptive antennas can employ, such as for enhancing the performance of mobile communications systems. The technique estimates and detects signals using their structural properties, but training sequences (known as 'bit fields') are used for initialisation of the estimation. The DILFAST algorithm combines array signals jointly in space and time. The algorithm performs separation, space-time equalisation and detection of multiple incoming signals. The received signal samples are mapped directly to the known *finite alphabet* (FA) constellation without any kind of subspace estimation. In addition to an underlying finite alphabet constellation of the modulation format, DILFAST utilises, optionally, the Toeplitz structure of the symbol matrix.

The algorithm is computationally very efficient leading to significant savings compared to subspace-based techniques [69]. The computational load of DILFAST is in the same range as conventional adaptive antenna techniques. Thus, the algorithm can be implemented in real-time with current base-station technology. A constraint is that the algorithm relies on the use of a finite and known alphabet, thus, for many scenarios this is practical, however, certain systems may change the alphabet due to variations in operating conditions.

4.16.2 Spectral Self-Coherence Restoral (SCORE) Algorithm

Most communication signals exhibit a property called *cyclostationarity*, which can be exploited to achieve blind adaptive beamforming. Cyclostationarity is a term used to describe the repetitive or cyclic nature of the statistics associated with signals. Periodicities in the second-order statistics of a signal leads to the existence of a correlation between the signal and the frequency-shifted and possibly conjugated versions of itself for certain discrete values of frequency shift. This property is called *spectral self-coherence* or *spectral conjugate self-coherence*. The cyclic correlation function and the cyclic conjugate-correlation function of a signal $x(t)$ are defined by equations (4.105) and (4.106) respectively.

$$R_{xx}^{\alpha}(\tau) = \left\langle x\left(t + \frac{\tau}{2}\right)x^*\left(t - \frac{\tau}{2}\right)e^{(-j2\pi\alpha t)} \right\rangle_{\infty} \tag{4.105}$$

$$R_{xx}^{\alpha}(\tau) = \left\langle x\left(t + \frac{\tau}{2}\right)x\left(t - \frac{\tau}{2}\right)e^{(-j2\pi\alpha t)} \right\rangle_{\infty} \tag{4.106}$$

where $\langle\,\rangle$ denotes the time averaging operator (or expectation factor),

$$\langle\cdot\rangle = \lim_{T\to\infty}\frac{1}{T}\int_{-\frac{T}{2}}^{\frac{T}{2}}(\cdot)\,dt \tag{4.107}$$

and τ and α denote the time delay and frequency shift, respectively. Most communication signals exhibit spectral correlation at one or more frequency shifts (carrier frequency harmonics or the baud rate or chip rate, etc.).

Based on the cyclic correlation function and the cyclic conjugate-correlation function, a class of *Spectral self-coherence restoral* (SCORE) algorithms has been developed. There are three SCORE algorithms reported in the literature: the least-squares SCORE, cross-SCORE, and auto-SCORE algorithms [70]. However, these algorithms cannot separate multiple signals having the same cyclic feature, which is the case in a CDMA system. Some variations of the algorithms have also been proposed to increase the convergence speed and reduce the computational complexity [71].

4.16.3   Constant Modulus Algorithm (CMA)

Some communication signals, such as frequency-shift keying (FSK), and analogue FM signals have a constant envelope. This constant envelope may, however, be distorted when the signal is transmitted through the channel. The constant modulus algorithm (CMA) adjusts the weight vector of the adaptive array to minimise the variation of the envelope at the output of the array. After the algorithm converges, the array can steer a beam in the direction of the *signal of interest* (SOI), and nulls in the direction of the interference.

CMA is a gradient-based algorithm that works on the basis that the existence of interference causes fluctuation in the amplitude of the array output, which otherwise has a constant modulus [33]. It updates the weights by minimising the cost function:

$$J(k) = \frac{1}{2}E[(|y(k)|^2 - y_0^2)^2]$$
(4.108)

using the following equation:

$$\mathbf{W}(k+1) = \mathbf{W}(k) - \mu\mathbf{g}(\mathbf{W}(k))$$
(4.109)

where

$$y(k) = \mathbf{W}^{\mathbf{H}}(k)\mathbf{x}(k+1)$$
(4.110)

is the array output after the kth iteration, y_0 is the desired amplitude in the absence of interference, and $\mathbf{g}(\mathbf{w}(k))$ denotes an estimate of the gradient of the cost function. Similar to the well-known LMS algorithm, discussed earlier in this chapter, it uses an estimate of the gradient by replacing the true gradient with an instant value given by

$$\mathbf{g}(\mathbf{w}(k)) = 2e(k)\mathbf{x}(k+1)$$
(4.111)

where $e(k)$ is the error signal between the reference signal and the array output.

The weight update equation for this case becomes

$$\mathbf{w}(k+1) = \mathbf{w}(k) - 2\mu e(k)\mathbf{x}(k+1)$$
(4.112)

CMA is useful for eliminating correlated arrivals and is effective for constant modulated envelope signals such as GMSK, which is used in digital communications such as the GSM cellular system. The algorithm, however, is not appropriate for CDMA based systems because of the power control used in such systems which dynamically controls the transmitted power as the propagation environment changes. The use of CMA to separate co-channel FM signals blindly in mobile communications has also been investigated, i.e., time domain filtering. A variation of CMA, referred to as *differential CMA* [72], has inferior convergence characteristics compared to CMA but may be improved using DOA information to make it operative in beam space.

4.16.4   Least-Squares Despread Respread Multitarget Constant Modulus Algorithm (LS-DRMTCMA)

LS-DRMTCMA is a blind adaptive beamforming algorithm which combines the spreading signal of each user in a CDMA system and the constant modulus property of the transmitted signal to adapt the weight vector [73], $\mathbf{W}$. In the base station of a CDMA system, the spreading signals of all the users are known beforehand.

If the kth data bit of the ith user is detected correctly, the waveform of the ith user's transmitted signal during time period $[(k-1)T_b; kT_b]$ can be obtained by respreading the detected data bit with the pseudo random sequence of the ith user, $ci(t)$. This respread signal can be used in the beamformer to adapt the weight vector for user i. The adaptive algorithm that uses this despread-and-respread and constant modulus technique is referred to as *Least-Squares Despread Respread Multi-Target Constant Modulus Algorithm* (LS-DRMTCMA).

4.17   CHAPTER SUMMARY

This chapter has presented a number of beamforming techniques which are broadly classified as:

- temporal reference;

- spatial reference;

- blind.

All the techniques aim to compute the weight vector, $\mathbf{W}$, such that the signal at the output of the beamformer in enhanced compared to those arriving at the array input. A number of algorithms have been presented along with the associated advantages and disadvantages.

It has not been possible to present all the variations of the algorithms and related performance results due to the quantity of information available, so the interested reader is referred to the references used throughout and to the open literature for such information.

4.18 PROBLEMS

Problem 1. Given that $L = 50$, $K = 10$ and $T_s = 0.5$ ms, use equation (4.4) to compute the convergence time of the array.

Problem 2. Assuming a two-element array with $d = \frac{\lambda}{2}$ that is to be used as part of a radar system for searching for new targets, compute the five weight vectors that will enable the beam to scan from $-20°$ to $+20°$ in $10°$ increments. Reference to main beam steering in chapter 2 may help.

Problem 3. Define what is meant by *cusping loss* in the context of switch beam beamforming and estimate the 3 dB beamwidth of an eight-element array with the main beam steered towards $+45°$, $d = \frac{\lambda}{2}$ and operating at 1.9 GHz .

Problem 4. Explain why an optimum beamformer does not require array calibration and state how, if calibration is available, it will enhance the operation of such a beamformer.

Problem 5. Derive the Wiener-Hopf equation (equation (4.22)). What is the main advantage of the method of steepest-decent compared to directly computing the Wiener solution?

Problem 6. How and why does employing a larger array improve the Fourier method of DoA estimation?

Problem 7. List the types of spatial smoothing that can be applied to adaptive beamformers. What are the main advantages and disadvantages of applying spatial smoothing?

Problem 8. List the primary advantages and disadvantages of blind beamforming algorithms.

5

Practical Considerations

5.1 INTRODUCTION

The concepts and algorithms presented so far have assumed idealistic conditions, i.e.,

- perfect beam weights are possible to compute and apply;

- channel estimates are known with a high degree of accuracy; and

- components do not cause signal distortion.

This is far from a realistic situation where radiowave propagation makes channel estimation a challenge in many operating environments and components cannot be manufactured to be consistent or have perfect parameters.

This chapter highlights many of the challenges a systems designer will encounter when implementing an antenna array and associated signal processing. In this chapter, techniques are introduced to overcome some of the challenges. A mobile communications system is used to provide a framework for the discussion and many of the techniques described have been developed as a result of the growth of mobile communications.

This chapter is organised as follows. Firstly, signal processing constraints are defined. A number of implementation issues are discussed, such as system calibration. Radiowave propagation and modelling are introduced, especially with respect to spatial propagation parameters and frequency effects. Finally,

Adaptive Array Systems B. Allen and M. Ghavami
© 2005 John Wiley & Sons, Ltd ISBN 0-470-86189-4

in contrast to the contents of the book so far which has been on signal reception, a number of transmit beamforming techniques are discussed.

5.2 SIGNAL PROCESSING CONSTRAINTS

In previous chapters, signal processing techniques have been introduced with a wide range of functionalities, such as finding of the DoA of a desired signal or cancellation of interfering signals. The performance and applicability of these algorithms have been gauged assuming provision of error-free input parameters, uncorrelated signals and distortion-free analogue components. Clearly, these conditions are not given in a realistic system and the question arises how much does the performance of aforementioned algorithms deteriorate as we diverge from the ideal case.

To answer this, algorithms have to be analytically, numerically or practically tested to see how sensitive they are to input errors or, as often referred to, to perturbations of the input parameters, one of the most important of which are the array weighting coefficients. As such, beamforming schemes, DoA algorithms and other adaptive algorithms rely in one form or another on a perfect estimate of these. The generally complex array weights, however, are prone to the following cause of errors [33]:

- computational errors due to finite-precision arithmetic;

- implementation errors induced by component variations;

- errors in the reference signal used to calculate weights;

- uncertainty in the exact position of each antenna array element.

These are known to cause random and hence unpredictable fluctuations in the weight amplitudes and phases. This in turn has a serious effect on the array gain, the sidelobe levels, the reduction in null-depth and the reduction in interference rejection capabilities. The most important sources of error and their impact on the system performance are summarised below.

5.2.1 Phase Error

An error in phase has a serious impact on any algorithm designed to estimate the DoA or to steer the beam in a given direction. A large error is known to lead to the suppression of the desired signal, where errors typically occur due to imperfect phase shifters and quantisation errors. The effect of a phase error is to steer the beams and nulls away from the desired angles.

5.2.2 Element Position Error

If an array beam pattern is achieved by means of constrained beamforming, then the induced steering vector error leads to a disturbed beam pattern and hence a diminished array gain. The element position uncertainty is known to produce a superposition of the intended pattern and that of a single element which results from the ideal and actual steering vectors differing.

5.2.3 Element Failure

A fault in the hardware or power supply may cause antenna elements to fail. That has obviously serious implications and system design should endeavour to minimise the risk of this happening. With fewer elements available, the beamwidth increases, and so does the sidelobe level. Element failure also reduces the available degrees of freedom, and the phasing of the feeding currents has to be recalculated taking the failure into account, otherwise the system operates sub-optimally and instabilities may occur. This may include unexpected grating lobes appearing.

5.2.4 Steering Vector Error

Ideally, the steering vector points in the direction of the desired signal. Algorithms generally rely on the fact that anything diverging from this direction is interference and is supposed to be suppressed or cancelled. Clearly, an error in the pointing direction leads the algorithm to believe that the desired signal is interference. This has serious implications, as the desired signal is being attenuated where attenuation increases with increasing pointing error.

Since steering vector errors are very common in antenna arrays, numerous counter-measurements have been devised. A simple but effective method is to broaden the beam and hence diminish the likelihood that the desired user falls outside the region where it is treated as interference.

5.2.5 Ill-Conditioned Signal Processing Matrices

From previous chapters, we can infer that almost every algorithm relies in one way or another on the inversion of system-related matrices. If these matrices become ill-conditioned, i.e., the ratio between largest and smallest eigenvalues approaches infinity, then an inversion becomes imprecise and unstable with finite-precision resolution. To overcome this difficulty, several methods have been developed which restore the rank of the involved matrices.

A typical example where ill-conditioning occurs is when the desired and interference signal are correlated. Then, the correlation matrix $\mathbf{R}$, statistically describing both signals, approaches singularity.

5.2.6   Weight Jitter

Weight jitter, otherwise know as *loop noise*, is a consequence of an adaptive beamformer utilising noisy gradient estimates, i.e., noise is present even after convergence. In real applications, temporal averaging is often not possible due to the real time constraint, as indicated in example 4.1. The consequence of weight jitter will depend upon the noise variance and is therefore linked to the amount of averaging performed. It is potentially catastrophic, but realistically will cause small perturbations in the array pattern. The effect of this will depend upon the sensitivity of the modulation scheme employed and the location of the mobile station in the beam pattern, i.e., a mobile station located in a null will be more sensitive due to the narrow width of a null compared to the main beam as illustrated in example 2.2.

5.3   IMPLEMENTATION ISSUES

Installation of antenna arrays for mobile communication systems (including GSM and third generation) and sonar, radar and biomedical applications involves the consideration of many factors, and ultimately compromises to meet both system performance and commercial cost imperatives. This section uses a mobile communications system to illustrate these factors. Physical constraints of a base-station (BS) antenna array are dependent upon the installation being considered. The three main types of BS installation are:

- *macro-cell*;

- *micro-cell*; and

- *pico-cell*.

Each are designed for different topographical and network usage. Typical examples of each of these installations are shown in figure 5.1. In each case a *synchronous digital hierarchy* (SDH) signalling system is transmitting using a microwave parabolic dish mounted below the array. This is used for communication between the BS and the cellular core network (often referred to as the *back haul*).

The macro-cell BS is designed to achieve maximum topographical coverage in semi-urban low to medium network usage areas. The antenna array in this case is usually mounted at heights of typically greater than 15 m to achieve maximum operating range, and minimise reflection and multi-path reception. The antenna array is typically located on a high building or purposely erected mast or tower that elevates it to above the average roof-top height. Micro-cell and pico-cell installations are designed for smaller topographical coverage within dense urban areas with high network usage. Micro-cell and pico-cell BSs are often physically constrained within a highly urban environment with buildings and other signal

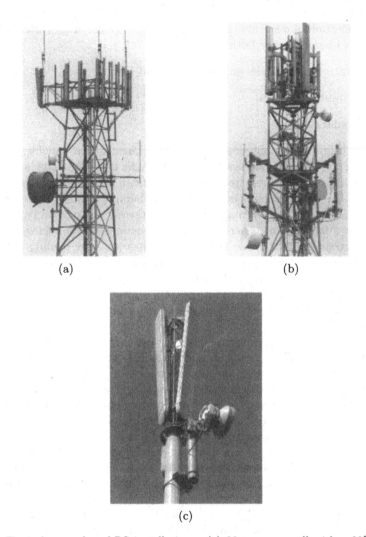

Fig. 5.1 Typical examples of BS installations. (a) 30 m macro-cell with a 30° sectored array; (b) 12 m micro-cell with multiple 120° sectored arrays; (c) 7 m pico-cell with a 120° sectored array.

"clutter". These base-stations are often deployed with a smaller array at reduced heights compared to macro-cell installations, for coverage of network 'hot spots'.

Deployment of a macro-cell BS antenna array has the most demanding implementation requirements of the three BS types and is considered here. Installation issues include:

- height requirements of the structure;

- selection of a suitable mounting for the antenna array (often a high building or mast/tower or both);

- static weight of the antenna array and feeders;

- wind-loading; and

- feeder selection.

The height of the antenna array is determined by the topology of its operating environment. GSM systems operate at 900 MHz and 1800 MHz, and 3G systems at 2.1 GHz. Signals at these frequencies experience high reflectivity, and exhibit poor penetration through obstructions such as buildings, and hills. Signal reception can be optimised by ensuring the height of the BS antenna array is commensurate with, or higher than, other physical obstructions within the system's coverage. The increased height of the antenna array increases the likelihood that the signal path between the *mobile station* (MS) and the BS will be *line of sight* (LOS). The additional height of the antenna array above ground level, however, has a direct effect upon the structure used for mounting the antenna array and feeder selection, which forms the next part of this discussion.

For the macro-cell BS antenna array, wind loading of the array and feeder loading (weight of the feeder cables) are two important structural considerations for installation. These result from long lengths of multiple feeders due to the array height, multiple elements, and the possibility of dual polarised elements. As an example, table 5.1 [74] compares the static weight required to safely stabilise an antenna with a 1 m^2 surface area onto a non-penetrating mounting (not fixed to a building or mast structure) at varying heights and wind loading. This surface area is similar to that required for a 900 MHz dual polarised sector antenna with a gain of 16 dBi. Note that this gain is achieved by employing a vertical panel of antenna elements connected together to achieve gain through increasing the directivity in the elevation plane. It can be seen that the height and environment have a significant impact upon the structural installation.

The selection of the signal feeder for the antenna array is an important issue. Table 5.2 [75] shows the attenuation characteristics for feeders typically employed in GSM and 3G BS installations.

Three factors are important in feeder selection for antenna array installation, these are:

- feeder loss;

- weight; and

- signal isolation.

Feeder loss increases linearly with length and therefore is related to the height of the antenna array. This loss is also frequency dependent, as can be seen from

Table 5.1 Static weight required to stabilise a 1 m² antenna on a non-penetrating mounting. Exposure (B): Antenna height of 9 m above ground, within an urban area with numerous closely spaced obstructions. Exposure (C): Antenna height of 15 m above ground, within open terrain, widely scattered obstructions generally less than 10 m in height.

Wind speed [kmph]	Total weight required [kg] for exposure (B)	Total weight required [kg] for exposure (C)
113	107	186
129	129	231
161	209	367
201	322	529

Table 5.2 Attenuation characteristics for LDF4-50A and LDF5-50A feeders.

Frequency [MHz]	LDF4-50A attenuation dB/100 m	LDF5-50A attenuation dB/100 m
400	4.46	2.49
600	5.53	3.10
800	6.46	3.63
1000	7.28	4.12
1250	8.23	4.67
1500	9.09	5.18
1800	10.1	5.75
2000	10.7	6.11
2300	11.5	6.63
3000	13.4	7.76

table 5.2. Figures 5.2(a) and 5.2(b) show the physical differences between LDF4-50A and LDF5-50A feeders. The figure illustrates the flexibility (bending radius) of the feeders, where LDF5-50A has little flexibility and is a relatively heavy cable but it is very robust and low loss. Feeder loss becomes important for maximum transmission power and minimum receiver noise figure requirements.

Transmission power from the BS will be attenuated by any feeder loss. Table 5.3 shows an example of a 100 W BS transmission at 2 GHz, fed through a feeder of 50 m in length. The attenuation and effective transmitting power for each feeder type is shown, for either case at least 50% of the transmission power is lost within the feeders.

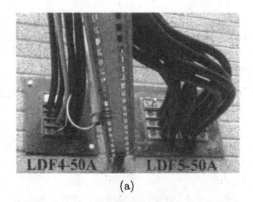

(a)

(b)

Fig. 5.2 Comparison of LDF4-50A and LDF5-50A feeders: (a) entering an equipment room routed from an antenna mast; and (b) fixed to a corner of a triangular mast section using fixing stays.

Table 5.3 Effective output power of a BS transmission via feeders.

Feeder type	Attenuation	Input power	Effective output power
LDF4-50A	5.35 dB	100 W	29.17 W
LDF5-50A	3.05 dB	100 W	49.54 W

For MS reception, the noise figure of the BS receiver is directly affected by feeder loss. The noise figure of a BS receiver can be defined as [76]

$$F_{\mathrm{RX}} = F_1 + \frac{F_2 - 1}{G_1} + \frac{F_3 - 1}{G_1 G_1} + \cdots \qquad (5.1)$$

where $F_1, F_2, \ldots$ are the noise figures of each successive stage within the receiver and $G_1, G_2, \ldots$ are the gains of each stage. It can be seen that the first stage (F_1) is predominant in setting the noise figure of the BS receiver. The noise figures of later stages are divided by stage gains which would be typically 30 dB, and therefore are much less significant.

The first components within the BS receiver are assumed to be the RF band pass filter (BPF), and low noise amplifier (LNA) (although it is possible to place these components alongside the antenna array, which, although this is desirable to reduce the accumulated receiver noise, it will increase the weight and hence wind loading of the mast and also pose problems should units require replacement due to failure). The noise figure of the first stage (F_1) is the sum of the insertion loss of the BPF, the noise figure of the LNA, and any other losses preceding the LNA which includes the feeder loss, as shown below

$$F_1 = f_{\text{loss}} + \text{BPF}_{\text{loss}} + F_{\text{LNA}} \tag{5.2}$$

where f_{loss} is the feeder loss, BPF_{loss} is the insertion loss of the BPF, and F_{LNA} is the noise figure of the LNA.

Inserting equation (5.2) into equation (5.1) gives

$$F_{\text{RX}} = f_{\text{loss}} + \text{BPF}_{\text{loss}} + F_{\text{LNA}} + \frac{F_2 - 1}{G_{\text{LNA}}} + \frac{F_3 - 1}{G_{\text{LNA}} G_1} + \cdots \tag{5.3}$$

It can be seen that the feeder loss, f_{loss}, proportionally increases the overall noise figure of the BS receiver. To maintain a low BS receiver NF, an LNA can be installed within a few metres of the antenna array, along with the Tx/Rx diplexer. The diplexer isolates the Tx and Rx signals during simultaneous transmission and reception when frequency division duplex is used, such as in GSM and 3G systems. The increase in noise figure due to feeder loss is reduced to

$$F_{\text{RX}} = \text{BPF}_{\text{loss}} + F_{\text{LNA}} + \frac{(f_{\text{loss}} + F_2) - 1}{G_{\text{LNA}}} + \frac{F_3 - 1}{G_{\text{LNA}} G_1} + \cdots \tag{5.4}$$

The feeder loss is added to the NF of the second stage. The feeder loss, which can be significant (3.05 dB and 5.35 dB in the previous example), is reduced by approximately the gain of the LNA which would typically be 30 dB. The noise figure of the LNA designed for operation on GSM or 3G systems is typically 0.5 dB.

Extending this idea, a down-converter can be connected to the output of each LNA near the antenna array. A separate feeder is required to supply the *local oscillator* (LO) frequency, however the down-converter allows much lower performance feeder cable to be used due to the lower frequencies it is required to carry. The cable is therefore both cheaper and lighter. As shown in table 5.2, feeder loss is frequency dependent, reduced performance cable will exhibit greater attenuation at higher operating frequencies, but this is addressed by the reduction of the input RF frequency to a much lower *intermediate frequency* (IF), by the frequency down-conversion process.

As an example of feeder considerations, from table 5.2, the LDF5-50A has typically half the attenuation of the LDF4-50A for a given frequency, however, it is significantly more expensive and has a much higher structural loading upon a mast. A 21 m high, 1.2 m wide triangular based three-section mast, with 12 lengths of LD5-50A feeder to the top, and 120° sectored arrays for operation at GSM 900/1800 MHz would fully load this mast. Such a configuration would support a twelve-element or six-element dual polarised array for micro-cellular deployment.

Future developments in antenna array technology look to address this issue of feeder loading and costs, using fibre optic cable or *Spatial Multiplexing of Local Elements* (SMILE) [77]. The conventional feeder cables can be replaced with a single very light and flexible fibre optic cable. This benefit is balanced by the need for additional hardware to be installed near the antenna array instead of in the equipment room at the base. For this to be realised small and lightweight hardware is required that has high reliability, as servicing and maintenance issues are more problematic to resolve at the top of the mast! The SMILE technique reduces the number of RF feeders required to one. The RF antenna inputs are time division multiplexed (TDM), then fed through the single RF feeder and de-multiplexed within the equipment room.

As deployment of micro-cells and pico-cells increases, particularly within 3G systems to fill network coverage gaps, and high network loading 'hotspots''', the aesthetic appearance of antenna arrays has come under scrutiny. Masts and arrays for third-generation mobile communication systems have been developed to be more aesthetically pleasing and sympathetic to the local surroundings. Installations have been developed for deployment within garage forecourt signs, street furniture including lampposts, telegraph poles, and artificial trees. Innovation is required with array topology to accommodate this relatively new aesthetic requirement. Figure 5.3(a) shows an example of a 3G BS disguised as an artificial tree, operational around the London M25 motorway in the UK. Circular antenna arrays have advantages for this application due to a reduced physical size. For a 3G pico-cell operating at 2.1 GHz, an eight-element circular antenna array with an antenna element radius of 0.5λ, has a diameter of only 15 cm and provides omni-steerable capability. The circular array may not be preferable for large multi-array installations due to the array dimensions restricting the physical mounting possibilities on an antenna mast, since increasing the number of elements results in an increased radius. The ideal position for the circular array is at the top of the mast to reduce scattering and mutual coupling from the mast structure that would otherwise impact the array characteristics and performance. The circular array may require a mounting pole through the centre of the array, alternatively the array can be mounted to one side of the mast.

Fig. 5.3 (a) A GSM micro-cell BS disguised as an artificial tree. (b) Base of GSM BS artificial tree, the feeder cables are routed within the tree trunk. (c) Top of GSM BS artificial tree, showing antennas mounted within the branches.

5.3.1 System Linearity

The transmit and receive radio modules in digital beamforming systems are required to be highly linear [78], i.e., the output of the system is a multiplication of the input as shown by equation (5.5). Conversely, equation (5.6) models a non-linear system containing higher order components resulting in signal distortion at the output.

$$y(t) = \alpha + \beta x(t) \qquad (5.5)$$

$$y(t) = \alpha + \beta x(t) + \gamma x(t)^2 + \delta x(t)^3 \tag{5.6}$$

The constraint on linearity is due to the weights being computed at baseband and subsequent distortion caused by the radio modules (up- and down-converters) will alter the phase and amplitude characteristics of the signals, resulting in a distorted radiation pattern. This is most prominent in the transmitter where the large signals require challenging circuit design in order to preserve linearity. Linearity impairments also occur at the receiver if the design does not have enough dynamic range.

Non-linear effects cause both *intermodulation distortion* (IMD) and amplitude and phase variations which fluctuate as the signal power is varied. Thus, these variations cannot be calibrated out in any way. Table 5.4 illustrates the impact in-band intermodulation products (IMP) have on sidelobe level (SLL), null depth (ND), main beamwidth (BW) and change in null direction (CND) for an eight-element $\lambda/2$ linear array employing Dolph-Chebyshev weights giving a 30 dB sidelobe level. The table [78] shows that the SLL and nulls alter considerably which will result in the spatial filtering properties of the adaptive array to be severely impaired. This can be seen by observing the reduced beamwidth and increased SLL, ND and large changes in CND as IMP are increased. Indeed, the CND alters by 7.4° for IMP = 17 dB and, with reference to example 2.3, this would result in a reduced nulling performance of 12.5 dB.

Table 5.4   Impact of IMP on the array radiation pattern.

IMP [dB]	BW [°]	SLL [dB]	ND [dB]	CND [°]
−17	13.3	12.6	29	7.4
−20	14.1	15	26.2	6.2
−27	15.6	21.1	24.6	2.6
−30	15.9	23.1	25.8	1.7
−75	16.4	29.6	42.5	0

5.3.2   Calibration

The spatial separation of antenna elements within an antenna array gives rise to magnitude and phase relationships between each antenna element when an incident plane wave impinges upon an array. Practical implementation of an adaptive antenna can distort these relationships if this distortion is not addressed by the use of calibration, and the performance of the adaptive antenna can be significantly reduced [79].

These errors, caused by the practical implementation of an adaptive antenna, are called 'calibration errors'. Calibration errors can distort the phase and magnitude relationships between the spatially separated antennas due to signals impinging on the array, and therefore give erroneous outputs. Calibration error in an adaptive antenna is caused by differences in component performance between different channels of the physical system. This is the subject of the next discussion.

Physical components are manufactured within specific tolerances. The manufacturing materials and construction of components used in the analogue RF and sampling components of an adaptive antenna make it impossible to produce them electrically 'identical'. Hence calibration is required to equalise the differences between the circuitry associated with the array elements (including the elements themselves). Figure 5.4 shows the analogue components used within a typical down-converter as part of an adaptive antenna BS receiver, i.e., between one of the antenna elements to the input of the *analogue to digital converter* (ADC). The difference between components used in the modules behind each array

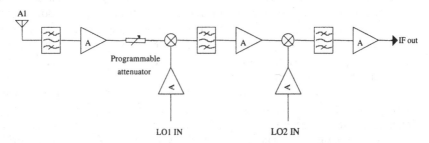

Fig. 5.4 Block diagram of a typical down-converter used within a BS.

element result in distortion to the signal path by addition of phase error, amplitude error, and noise. The transfer function for the signal path of each antenna element as it progresses through the system is different as each signal is distorted in a different way. This is illustrated in figure 5.5. The dominant types of distortion are:

- variation in group delay between filters;

- differences between amplifier gains;

- tolerance in attenuator accuracy;

- RF feeder construction; and

- aperture jitter on ADCs.

Aperture jitter is introduced at the sampling stage and is caused by small variations in the timing of the analogue signal being sampled by the ADC.

This is caused both by tolerances internally within the ADC itself and by the ADC input clock generation circuitry [80]. Unless the distortion is a known

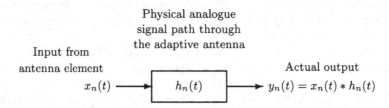

Fig. 5.5* Effect of signal distortion on the signal path for one antenna element.

quantity to allow compensation or calibration to be applied, spatial algorithms operating upon an adaptive array may be significantly affected, as the internal distortion to each signal path is uncorrelated with the signal. Consequently the uncorrelated distortion causes the effect of weighting each antenna input and offsetting the signals impinging upon the array. The distortion introduced by practical implementation of an adaptive array can be divided into two groups termed here as *static* and *dynamic*.

Static distortion accounts for differences in component performance due to manufacturing tolerance in the physical construction. These parameters, once calculated, are relatively stable over the life of the component, allowing for component ageing. Dynamic distortion is caused predominantly by thermal effects, but component ageing can also cause long-term variations. Components are specified to operate within a given tolerance, but performance varies within this range. This variation will also be different for apparently similar components. These dynamic errors are the most problematic and require the most attention since calibration must be performed frequently enough to track the associated variations. Within the TSUNAMI 2 smart antenna field trials [81], calibration measurements are taken within GSM time slots. Variation of the operating temperature of the adaptive antenna hardware is the most significant factor of dynamic distortion. From a comparative test of two down-converters constructed as shown in figure 5.4, the difference between the signal delay of the outputs for temperatures of $0°$ C, $25°$ C and $50°$ C have been obtained experimentally and plotted as a function of IF frequency. This is shown in figure 5.6. It can be seen that the signal delay difference between the outputs varies significantly with temperature. This affects the ability of the adaptive antenna to form the desired beam as the added delay from the down-converter appears to the signal processor as though the input plane wave originates from a different angle of azimuth. It can be seen that the variation in delay is greatly accentuated at frequencies away from the centre intermediate frequency of 70 MHz.

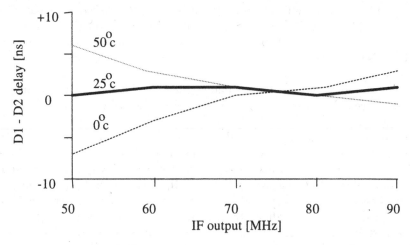

Fig. 5.6 Variation in signal delay between two down-converter outputs (D1 and D2) due to change in operating temperature.

Switched beam systems [82] are inherently limited to the number of beams that can be generated, and cannot provide dynamic control of individual beam patterns; however, the interdependence of the analogue hardware is reduced which offers reduced calibration requirements. These beams are generated by an analogue beamforming network at the RF input. The signals are then fed via down-converters for demodulation. As the down-converters are relatively independent, the effect of variation between their performance due to temperature is greatly reduced.

5.3.2.1 Adaptive antenna design for calibration. The purpose of a calibration process is to either correct the distortion before processing begins, called *pre-distortion* correction, or to calculate the distortion between all channels of the adaptive antenna so this can be used as correction data later, called *post-distortion* correction. Both have applications in practical systems.

As well as pre- and post-distortion correction techniques, *on-line* and *off-line* calibration is possible. Off-line calibration corrects static errors, this is a one-off calibration of the system that can take place in situ or in a laboratory. On-line calibration corrects for dynamic errors, due to temperature and component ageing. On-line calibration can also correct for static errors, but initial use of off-line calibration brings the system parameters to within manageable bounds of a practical on-line calibration system, therefore enabling the design parameters of the on-line calibration system to be relaxed. It is important to note that optimal beamformers adapt weights until optimal conditions are met; however, the calibration error may exceed a signal wavelength. This is exacerbated as the operating frequency is increased towards those often used for radar applications, i.e., shorter operating wavelengths will mean even relatively small errors will exceed a wavelength. Applying calibration will bring the adaptive antenna within bounds to reduce the optimal beamformer's search range and time to reach optimal performance.

There are several design techniques that can be employed to reduce distortion introduced by the system design and consequently enable the calibration requirements to be relaxed [83]. Conventionally the I/Q (in-phase and quadrature) channels are generated by analogue components and the baseband signals are then sampled by an ADC, this is shown in figure 5.7. The analogue mixers used to achieve this can introduce phase errors of typically between 1° and 3° and amplitude unbalance of up to 0.5 dB between the I/Q channels [84]. The 90° phase splitter also introduces errors. To overcome this the IF input can be sampled directly, this allows the use of *digital down conversion* (DDC) [80]. The IF is sampled and I/Q generation is replicated in the digital domain without the introduction of analogue component errors.

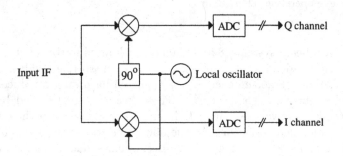

Fig. 5.7 Analogue I/Q generation.

Sampling at frequencies higher than baseband can reduce the dynamic range requirement of the system when implementing the design. At baseband data rates, ADC resolution is comparable to the dynamic range of the analogue mixers, so linearity of the system is not reduced. A limitation of sampling at the IF is the degradation of sampling resolution of the ADCs. Due to hardware limitations, as the sampling frequency of ADCs increases, the bit resolution (and therefore dynamic range) of components reduces.

As an example, consider high specification devices for a given resolution, in the Analog Devices range of ADCs, the AD7674 is an 18 bit device with a sample frequency of 800 ksps, the AD6645-105 is 14 bit device with a sample frequency of 105 Msps. The reduction in bit resolution reduces the dynamic range of the system, in this example theoretically reduced from 108 dB for the 18 bit ADC to 84 dB for the 14 bit ADC, a reduction of 24 dB, which is significant and affects operation in situations where the received signal can vary considerably.

This should be considered within the design of the system. The number of ADCs required in sampling the IF, however, is halved when compared to sampling the baseband I and Q signals generated as in figure 5.7, since complex sampling (I and Q) is not required since this processing is subsequently completed in the digital domain.

The process of *digital down-conversion* (DDC) can be extended to provide further implementation advantages using harmonic sampling [80]. The use of DDC at the IF allows digital generation of the I and Q channels without the magnitude and phase error introduced by the analogue components used in baseband sampling. This implementation has been shown to reduce the dynamic range of the system due to hardware limitations of the ADCs. This is exacerbated as the sampling frequency is increased to sample higher frequency IFs. The digital signal processing (DSP) requirements for post sampling are also significantly more complex and costly due to the required processing speeds. These increase from several thousand floating operations per second (kFLOPS) for baseband sampling to millions of FLOPS (MFLOPS) for IF sampling using DDC.

Harmonic sampling allows the sampling frequency to be reduced lower than the input IF. This allows the dynamic range of the system to be maintained at the same time allowing digital generation of the I/Q signals. Many ADC devices are now available to achieve this. Harmonic sampling appears to break the Nyquist sampling theorem of using a sampling frequency of at least twice the maximum input frequency to avoid aliasing. Harmonic sampling can be used only if certain signal and sampling criteria are met. The key criteria are that:

- the IF is suitably *bandpass filtered* (BPF);

- multiples of the sample frequency do not appear within the IF signal bandwidth;

- the analogue section of the ADC extends to the IF; and

- the aperture jitter of the ADC is commensurate with the IF and not the sampling frequency.

For example an input IF of 60 MHz with bandwidth of 10 MHz could be harmonically sampled with a 10 Msamples/s sample clock if appropriate filtering was applied, and ADC was selected.

The bandwidth of the system can determine what type of calibration is applied to the system. Narrowband systems typically have a fractional bandwidth (FB) of less than 1%, wideband systems typically a FB greater than 1%. *Narrowband calibration* calibrates the centre frequency of system operation. *Wideband calibration* is applied to calibrate over the whole frequency range of the system and can be achieved using adaptive filters; this technique applies calibration factors to each sample frequency.

Narrowband calibration is the simplest form to implement practically. Calibration of the signal magnitude can be achieved by taking the average power of the sampled signals and applying a compensation multiplier. Calibration of signal phase can be achieved by adjustment of the sample clock, this is commonly referred to as *sample clock dither* (SCD) and is illustrated in figure 5.8. The range

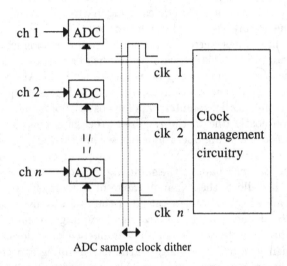

ADC sample clock dither

Fig. 5.8 Example of sample clock dither.

of adjustment required using SCD is important when considering a *continuous wave* (CW) source for calibration purposes. For a standard 10.7 MHz IF, the period of the wave is 93.46 ns. This value can be comparable to variation in delay due to component tolerance and temperature effects. Using a CW source, variation greater than one period cannot be distinguished, which poses a problem in achieving optimal weights in optimal beamforming within an acceptable time window. To overcome this, a modulated source can be used as a test burst. The known modulation pattern provides a time line reference against which relative movements of the centre frequency on any channel can be compared against.

To calibrate the phase of each signal using SCD, the resolution of a master clock will be limited for fine phase adjustment. As an example, consider a 40 Msamples/s sample clock with an IF of 10.7 MHz. In this case each sample interval represents 96.3° of phase shift, which will clearly yield an unacceptable performance.

For finer control of phase adjustment *digital clock management* (DCM) can be used. For hardware implementation of clock control circuitry programmable logic devices or DSP's offer clock multipliers with digital phase control. Typically, fractional adjustment using these devices is $\pm 255^{\text{th}}$ of the master clock. In this case the resolution of the phase adjustment using sample clock dither would reduce from 96.3° to $\pm 0.377°$.

A technique called *Spatial Multiplexing of Local Elements* (SMILE) has been demonstrated in [77] to reduce mutual coupling between antenna elements in an antenna array, and also to reduce hardware complexity of the system. Figure 5.9 demonstrates an adaptive antenna using SMILE. This technique can also be applied to reduce the calibration requirements of an adaptive antenna system. This is achieved by reducing both the analogue hardware and samplers, and hence the effects of variation of these relative to each signal channel with temperature, ageing, and sampling jitter etc. The SMILE technique time-division multiplexes

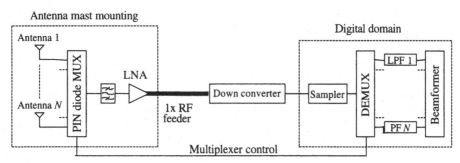

Fig. 5.9 Diagram of SMILE implementation.

(TDM) the received signals at the antennas. After multiplexing, the signals are fed through a single RF feeder and down-converter. The output of the down-converter is sampled, and all further processing including de-multiplexing of the signals is implemented in the digital domain. Note that this architecture requires the bandwidth of the down-converter and the sampling frequency to be N times higher than that of a conventional parallel architecture. The relative distortion between each signal channel introduced by variation between down-converter outputs (figure 5.6) is eliminated. All channels are now subjected to the same distortion.

The benefits of this technique are balanced with several implementation issues. The TDM of the signals at the multiplexer output increases the bandwidth of the RF signal by a factor equal to the number of antennas. The down-converter bandwidth and sampling frequency hence have to be similarly increased. The multiplexer will introduce signal loss and therefore increase the noise figure of the system. Also, the TDM of the input signals by the multiplexer will introduce switching noise, reduce signal isolation between the channels and may introduce calibration issues itself. Many of these issues, however, can be overcome with careful system design, although this technique is currently in its infancy. It can be seen that calibration of an adaptive antenna is an important issue for correct operation of adaptive antenna systems and even plays a role in improving the operation of optimal combining beamformers which are often perceived not to require calibration. Thermal effects are one of the main contributors to calibration errors in a practical system, the temperature dependence has also been shown to be frequency dependent. The question of whether to employ narrowband or wideband calibration should therefore be carefully considered when implementing a practical system. Narrowband calibration is considerably simpler to implement. As temperature-controlled environments are very difficult to achieve, especially for mobile equipment, regular automatic calibration of the system is required to maintain operating performance. Several design techniques have been discussed to reduce inherent calibration errors in the design of an adaptive antenna system including DDC, HS, SCD and SMILE.

5.3.3   Mutual Coupling

The analysis of antenna arrays so far has assumed perfect elements which act independently of each other. The performance of an adaptive antenna can be affected by interactions between individual elements of the antenna array. This interaction is caused by electromagnetic coupling between the antenna elements, and is commonly known as *mutual coupling* [85]. As a result of mutual coupling the gain and radiation pattern of an isolated antenna element may be significantly different when used with the antenna array. The array may also be subject to scattering of signals caused by the array mounting hardware.

The extent of mutual coupling acting upon an antenna element can be dependent upon its location within the array, and distance between each element. Antenna elements at the edge of the array will be affected differently to antennas in the centre of the array. The *Uniform Circular Array* (UCA) has benefits over the *Uniform Linear Arrays* (ULA) in respect of mutual coupling. Due to the symmetry of the UCA the effect of mutual coupling becomes periodic and symmetric around the array, whereas elements at the ends of the ULA will have different mutual coupling characteristics from the centre elements.

Given a plane wave incident upon the antenna array, the effect of mutual coupling on the radiation pattern of the antenna elements within the array is derived from the direct re-radiation from each element itself and additional re-radiation from antenna elements illuminated by the incident signal [85]. Figure 5.10 depicts an incident signal with re-radiated signals transmitted to element one of an eight-element UCA. The result of mutual coupling on the beams formed by an adaptive antenna is increased sidelobe powers, i.e., reduced sidelobe levels. This affects the ability of the beamformer to distinguish the DoA of signals, and to adapt the beams for best signal to interference plus noise ratio (SINR). Several techniques can be applied to the adaptive antenna either to reduce the effects of mutual coupling on the adaptive antenna by using signal processing, or to reduce the mutual coupling in the physical array. This is the subject of the following discussion.

It can be shown that a typical reduction of 3 dB in sidelobe power can be achieved when mutual coupling is taken into consideration, also, that for a UCA

Retransmitted signal to element 1

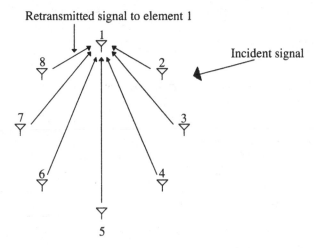

Incident signal

Fig. 5.10 Re-radiation of signals from the incident plane wave to element 1 of the UCA.

the mutual coupling experienced is both periodic and symmetric [86, 87]. These results also imply that mutual coupling compensation has less effect at signal with direction of arrivals that match the location of the array elements, and the effect of scattering from one element to another is minimised when the DoA matches the angle of the antenna element in the array with respect to the array centre.

Off-line calibration can be used to reduce the effect of mutual coupling on the adaptive antenna. This reduces the sidelobe power of the beamformer, and also improves direction of arrival estimation without the need for field calibration. Off-line calibration is based upon consideration of the antenna elements and their geometry upon a fixed structure; the mutual coupling can then be assumed to be non-changing within the system operation. The mutual coupling can be calculated by using the numerical electromagnetics code (NEC) [87].

Mutual coupling perturbs the array performance, which impacts DoA algorithms such as the MUSIC subspace method [88] and as described in chapter 4. DoA applications are being demanded in modern communications networks such as GSM and 3G for the use of *location based services* (LBS), where the network can provide users with location information such as nearest railway station etc. This is described further in the next chapter.

Simulations undertaken in [89] apply the MUSIC DoA algorithm to an adaptive antenna with a seven-element circular array, and compares the performance with and without mutual coupling compensation. The mutual coupling is calculated from what is commonly known as a *coupling matrix* which describes the amount of coupling between each element, including itself (self-coupling) which assumes a value of 1. By measuring the actual response to a few known incident angles during calibration, the coupling matrix can be estimated. A summary of the results of several trials is shown in table 5.5. It can be seen that the compensation provided by calculation of the coupling matrix gives correct DoA estimation, whereas this is not correct in most cases when compensation is not used. By reducing the radius

Table 5.5   Results of simulations of DoA with and without compensation applied.

	Array radius	Correct DoA uncompensated	Correct DoA compensated
Single incident wave	$\frac{\lambda}{2}$	Yes	Yes
Incident signals 30° and 50°	$\frac{\lambda}{2}$	No	Yes
Change in antenna impedances	$\frac{\lambda}{4}$	No	Yes
Array perturbed by mounting pole in the centre of array, radius $\frac{\lambda}{5}$	$\frac{\lambda}{2}$	No	Yes

of the array from $\lambda/2$ to $\lambda/4$, mutual coupling in the array is increased. Where near-field scatterers (mast, array mounting etc.) are $\lambda/2$ or greater away from the antennas, mutual coupling can be shown to become negligible [90].

The previous techniques have been aimed at calculation of mutual coupling to apply compensation. Another technique called spatial multiplexing of local elements (SMILE) [77] (as described in the previous section and illustrated in figure 5.9), can reduce mutual coupling to that approaching the model for a single antenna element. This is achieved as only one antenna in the array is terminated in a load impedance of 50 Ω at any one time, all others are either open or closed circuit depending on the frequency. This is shown in figure 5.11. However, a drawback with this system is that the insertion loss of the multiplexer directly adds to the noise figure of the system. Mutual coupling has an adverse effect upon the performance of an adaptive antenna by increasing the sidelobe powers. This in turn affects the ability of the adaptive antenna to estimate the DoA of signals. It is possible to calculate the mutual coupling and apply this as a compensation factor to reduce the effect of mutual coupling. Special design techniques can also be used to reduce the mutual coupling in the array, such as SMILE.

5.3.4   Circular Arrays

Circular arrays have received increasing attention for application in pico-cell BSs [91]. The omni-steerable capability in azimuth and compact topology compared to linear arrays make it applicable to small aesthetically acceptable installations such as these. A plane wave impinging upon a circular array can be represented in three-dimensional space by figure 5.12, and by the following expression

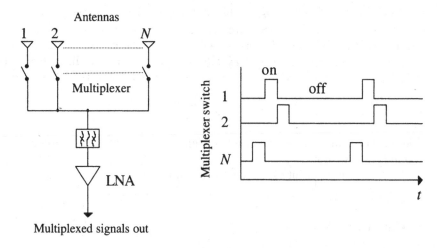

Fig. 5.11 Example SMILE switching circuit.

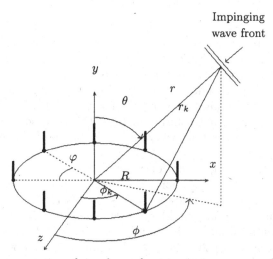

Fig. 5.12 Representation of signal wavefront impinging on a circular array.

$$\beta_k = \kappa R \cos(\Phi - \Phi_k) \sin \theta \tag{5.7}$$

where β_k is the relative phase of array element k to the centre of the array, κ is equal to $2\pi/\lambda$, λ is the wavelength, ϕ is the azimuth angle of the signal impinging on the array and R is the array radius.

Given the increased interest in the deployment of circular arrays it is interesting to outline comparisons of the circular array particularly with the predominant alternative, the linear array. Here we will limit the discussion to the most common types of the uniform circular array (UCA) and the uniform linear array (ULA), which have equi-spaced antenna elements. In figure 5.12 for the UCA we can write

$$\varphi = \frac{360}{N} \tag{5.8}$$

where φ is the angle of separation between each antenna, and N is the number of antennas in the array.

The parameters of interest for practical adaptive antennas include:

- beam steering;

- sidelobe levels;

- interference rejection; and

- application of signal processing algorithms.

In particular, algorithms for estimation of the direction of arrival (DoA) of signals have been predominantly applied to ULAs due to the Vandermonde structure of this type of array. A *Vandermonde structure* refers to progressive linear phase shift of the signals across certain array structures such as the ULA. This lends itself to concise mathematical derivation. The UCA has comparably larger sidelobes than a ULA with an equivalent number of antenna elements [35]. Unlike the ULA, however, the angular resolution of the UCA is reasonably well maintained over the full 360° of operation, whereas the usable beamsteering capability of a ULA is typically 120°.

To reduce the higher sidelobe powers within the UCA beam-pattern, spatial windowing [92] with an appropriate length spatial filter can be applied to the azimuth response (see chapter 2). Using this technique, spatial windowing can provide narrower beams and lower sidelobe designs for the UCA. Examples of the types of spatial windows that can be applied include binomial, Dolph-Chebyshev and Taylor-Kaiser [92]. The Dolph-Chebyshev window provides the optimum solution. It maintains all sidelobe powers at an equal level, and for a given sidelobe power, it produces the narrowest beam width (see chapter 2).

An adaptive antenna employing a UCA can outperform ULAs at rejecting unwanted signals when an equivalent number of antenna elements are used for omni-steerable capability [93, 94]. Reference [93] compares a twelve-element UCA and three four-element ULAs arranged in a triangle to provide 360° operation. The UCA outperforms the ULA array at rejecting unwanted signals; this is due to the twelve-element UCA having more degrees of freedom than the three separate ULAs. Conceptually, if interference is within one of the ULA 120° sectors, the benefit of more degrees of freedom produced by the UCA can be brought to bear to produce a larger number or deeper amplitude nulls in the azimuth beam shape. This remains valid even when considering that interference in one of the ULA arrays can be reduced by a factor of three by ignoring the signals from the other two ULA arrays.

Much research has been undertaken into signal processing algorithms applied to adaptive antennas with ULAs, particularly for estimation of signal DoA. These algorithms include the multiple signal classification algorithm (MUSIC) [88], as described in chapter 4. It would be advantageous to be able to apply this body of research to the UCA. One method to achieve this is to use a translation technique to convert the UCA to a ULA response. Such a technique is the *Davies transformation* [95, 96], here the UCA is transformed to a virtual ULA by the Davis transformation (**T**); this is depicted in figure 5.13. There are several

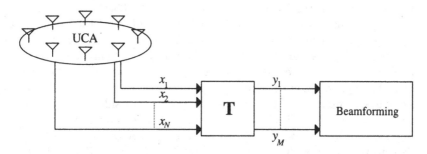

Fig. 5.13 Representation of the Davies transformation to translate a UCA response to a virtual ULA response.

caveats that have to be applied to make the Davies transformation accurate. One of these is that the number of antennas in the UCA has to be high enough to approximate a constant array pattern over all steering angles. The *shading matrix* (i.e., amplitude weighting as described in chapter 2) that is subsequently applied to the array pattern is then independent of the steering angle. This ensures the required Vandermonde structure of the virtual ULA response. The closed form solution for the Davies array can be shown to be as follows [91]: If $\mathbf{b}(\theta)$ is the steering vector of the Davies array and $\mathbf{a}(\theta)$ is the steering vector of the UCA

$$\mathbf{b}(\theta) = \mathbf{T}_{D_{\text{av}}}\mathbf{a}(\theta)$$

$$\left[\mathbf{b}(\theta)\mathbf{a}^{H}(\theta)\right] = \cos\left[\eta\theta + kr\sin(\theta + \gamma)\right]c - j\sin\left[\eta\theta + kr\sin(\theta + \gamma)\right] \quad (5.9)$$

where

$$\lambda = \frac{-2\pi(q-1)}{N} + \frac{\pi}{2}$$
$$\eta = M_0 + 1 - p$$
$$p = -M_0, \ldots, 0, \ldots, M_0$$
$$q = 1, \ldots, N$$
$$M = \text{number of virtual ULA elements}$$
$$N = \text{number of UCA elements}$$

The Davies transformation can be shown to be highly susceptible to even small permutations in the array response [97]. This can have a serious effect on the performance of DoA estimation algorithms such as MUSIC. This can be resolved by monitoring large value errors appearing in the Davies transformation and applying further signal processing to reduce these errors. These approximation errors in the transformation of the array response are traded for the required vector with a Vandermode form to give robustness. A perturbed UCA array transformed to a ULA using this modified Davies transformation, can be shown to give DoA performance using the MUSIC DoA algorithm that approaches an ideal UCA [97]. A simple circular array model has been introduced. The UCA has increased sidelobes compared to an equivalent ULA. These sidelobes can be reduced by spatial windowing. The UCA has been shown to provide improved interference rejection for an equivalent number of antennas. A Davies transformation can be used to transform a UCA to a virtual ULA to allow algorithms developed for the ULA to be used. This transformation is essentially a spatial DFT.

5.4   RADIOWAVE PROPAGATION

So far the discussion has focused on antenna elements, arrays and related signal processing. However, understanding the nature of the signals impinging on the arrays which sets fundamental requirements on the design of the subsequent antennas, transceivers and activated algorithms is paramount to ensuring a successful performance. This section describes channel modelling and radiowave propagation characteristics that relate to the operation of antenna arrays, and is presented in four parts. Firstly, the single antenna narrowband channel model is reviewed. This is then extended for the case of multiple antennas where the spatial domain should be included in the model. Following this, the multiple antenna channel model is then enhanced to include the channel impulse response which occurs when transmissions are wideband. The terms narrow- and wide-band are defined in this context to be transmissions with bandwidths that are either less than or greater than the coherence bandwidth of the channel under consideration (this is similar to comparing the symbol duration of a digital communication system with the normalised channel delay spread). The coherence bandwidth, B_c, is computed from the complex frequency response of the channel, $H(f)$. The magnitude of the correlation coefficient, $|\rho|$ is first computed from the channel's frequency response for each pair of samples and is computed using [98]:

$$|\rho_{11}| = \left| \frac{R_{ij}}{\frac{1}{N}\sqrt{\left[\sum_{i=0}^{N-1} x_i^2(f_n) \cdot \sum_{j=0}^{N-1} x_j^2(f_n)\right]}} \right| \qquad (5.10)$$

where R_{ij} is the cross-correlation for zero lag, $x_i(f)$ and $x_i(f)$ are the ith and j^{th} complex frequency response samples and N is the number of independent samples of $H(f)$. Finally, B_c is defined as the frequency separation at which the correlation coefficient falls below some threshold, ρ_{th}. $\rho_{th} = 0.5$, 0.7 and 0.9 are commonly chosen thresholds. The choice of ρ_{th} will depend upon the amount of frequency selective fading the system is desired to operate with. Higher values of ρ_{th} yield a lower coherence bandwidth, consequently the system operating within this bandwidth will be subject to less frequency selective fading than if it was operating at a wider bandwidth specified if a lower value of ρ_{th} was used. Figure 5.15 shows an example channel frequency response, $|H(f)|$, and its associated correlation function, $R(f)$, is shown in figure 5.16. The figure was obtained by measurement over a 20 MHz bandwidth at 2.1 GHz. The equipment was placed in an urban environment with the mobile station on the road and the base station above rooftop height as described in [99]. The following values of B_c are obtained for correlation thresholds of 0.5 and 0.9:

- $B_{c0.5}$ =1.2 MHz

- $B_{c0.9}$ =0.31 MHz.

This section illustrates how the spatial channel changes with frequency. This is important when an adaptive antenna has to operate on two (or more) frequency bands (which is the case when simultaneous transmission and reception occurs on different frequencies - referred to as *frequency division duplex* or *FDD*), or when it is required to operate over a very wide bandwidth. GSM and third-generation mobile communication systems both employ FDD.

Although this section focuses on wireless channels, i.e., communications systems and radar, the underlying principles can be applied to sonar systems where wave propagation occurs in water. The only difference being that the material characteristics of the propagation medium will differ accordingly i.e. transmission through water at varying temperatures, pressures and salinity.

5.4.1 Narrowband Single Antenna Channel Model

In chapter 2, the array steering vector, **a**, was shown to model the signals arriving at the array elements. In a practical scenario the signals arriving at each element will undergo attenuation, reflection and diffraction as they propagate from the transmitter to the receiver. On arrival at the receiver, these signals are then subject to the array steering vector. Thus signals arriving at the array, $\mathbf{U}$, need to be modelled. The signals arriving at the elements will also have some degree of correlation between them which must also be accounted for in the model.

When a single antenna is used at each end of the link the signal will be distorted by the following factors:

- gain of transmit antenna (G_T);

- distance dependent path loss (L_p);

- log-normal shadowing (s);

- Rayleigh or Rician fading (ξ);

- phase lag (ψ); and

- gain of receive antenna (G_R).

The received signal, a, is a multiplication of these factors as given by the equation below.

$$U = G_T \cdot L_p \cdot s \cdot \xi \cdot \psi \cdot G_R \tag{5.11}$$

Numerical models for each of these factors are well known and are well documented in references such as [100, 101]. ψ is an element in the array steering vector and is a phase shift caused by the time of propagation, which is a function of the path length between the transmitter and receiver.

5.4.2   Multiple Antenna Channel Model

When an N element antenna array is used at the receiver, the received signal at the output of each element will have been subject to the above factors and a multiple antenna channel model is therefore required. It is therefore convenient to use vector notation, where the received signal vector, $\mathbf{U}$, is given by:

$$\mathbf{U} = [\ U_1 \quad U_2 \quad \dots \quad U_N \ ] \tag{5.12}$$

and each element of $\mathbf{U}$ is computed using equation (5.11). Each of the elements of $\mathbf{U}$ are assumed to be independently generated from a mathematical model, however, in a real propagation scenarios there will be a degree of correlation between the signals arriving at each element. This spatial correlation is described in a matrix, $\mathbf{R_s}$, as shown below for a two-element array.

$$\mathbf{R_s} = \begin{pmatrix} 1 & \rho_{12} \\ \rho_{21} & 1 \end{pmatrix} \tag{5.13}$$

where the diagonal represents auto-correlation of the elements and is obviously equal to 1, and the anti-diagonal elements represent the cross-correlation between the elements. $\mathbf{R_s}$ is referred to as the *spatial correlation matrix*. The values of ρ_{12} and ρ_{21} are determined by the element spacing, orientation, radiation pattern and nature of the propagation environment. $\mathbf{R_s}$ is incorporated into the channel model as:

$$\mathbf{U}_R = \sqrt{\mathbf{R}}\mathbf{U} \tag{5.14}$$

where $\mathbf{U}_R$ is the spatially correlated received signal vector and $\mathbf{U}$ is the uncorrelated received signal vector.

5.4.3   Wideband Multiple Antenna Channel Model

When the transmission is wideband, i.e., when the bandwidth of the signal is wider than the coherence bandwidth of the channel, B_c, where B_c can be computed using equation (5.10), the signal will undergo frequency selective fading which results in an impulse response of the channel being distinguishable in the time domain at the receiver. The channel impulse response is classically represented as [102]:

$$h(t,\tau) = \sum_{n=0}^{N(t)-1} \xi_n(t)e^{j\psi_n(t)}\delta(t - \tau_n(t)) \tag{5.15}$$

Where $N(t)$ is the number of resolvable multipath components, ξ_n is the amplitude of the lth multipath component at time t, φ is the carrier phase at time t and $\delta(t - \tau_n(t))$ represents a delay. An example impulse response measured in the centre of the city of Bristol, UK is shown in figure 5.14 [99] and represents the time domain of the channel's frequency response shown in figure 5.15. The measurement was taken whilst an eight-element antenna array was mounted on a rooftop and the mobile station was at street level, hence the impulse response for each of the eight channels are shown. The operating frequency was 2.1 GHz and the bandwidth used to produce the figure was 5 MHz, i.e., part of the total bandwidth used to produce figure 5.15. These operating parameters make the impulse response similar to that experienced by 3G UTRA cellular radio systems [99].

Due to the finite transmission bandwidth, not all of the multipath components are resolvable. This results in each of the components of the impulse response consisting of several signals, and results in Rayleigh or Rician fading of each component, i.e., if a wider bandwidth (say 1 GHz) was used, more impulses would be distinguished and almost no fading would occur. The figure shows the most energy arriving at 1.5 μs and decaying as the time delay is increased. The vertical axis is path loss, hence the signals arriving at 1.5 μs have experienced a path loss of around 80 dB. The figure also shows data for 8 channels, which is due to an eight-element linear array being used for these measurements. It can be seen that the path loss does not vary much across the eight elements. The above channel yields frequency selective fading since the signal bandwidth is 5 MHz and the coherence bandwidth is 1.4 MHz. This is confirmed by the existence of multiple impulses. Thus frequency selective fading is observed and is shown in figure 5.15. The figure shows the channel frequency response, $|H(f)|$, of this radio channel for each of the eight array elements and is computed by taking the Fourier transform of the channel impulse response of each element.

Note that the 1.5 μs delay of the first impulse arriving represents the *time of flight* (ToF) of the signal which enables the path length (l) to be computed from using:

$$l = c \cdot ToF \tag{5.16}$$

where c is the speed of propagation, i.e. 3×10^8 m/s for electromagnetic waves. Thus, $l = 450$ m for this example.

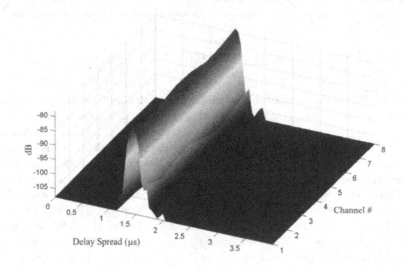

Fig. 5.14 Example impulse response for a 3G cellular system channel in an urban environment.

The above wideband channel model does not consider the angle of arrival of each multipath component. This is simply added by including the array steering vector, **a**, and yields the wideband multiple antenna channel model:

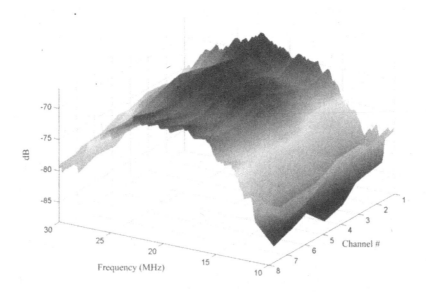

Fig. 5.15 Example frequency response of a 3G cellular system channel in an urban environment.

$$h(t,\tau) = \sum_{n=0}^{N(t)-1} \xi_n(t)e^{j\psi_n(t)}\mathbf{a}(\theta_n(t))\delta(t - \tau_n(t)) \tag{5.17}$$

It can be seen from the above equation that the steering vector is dependent upon delay. This is observed in figure 5.17 which shows the angle of arrival of signals as a function of delay, i.e., the figure depicts the impulse response as a function of angle. The measurement was taken in the same operating environment as that described above. The figure can be likened to a radar plot of the operating environment and is often referred to as a *scatter map*. The *directions of arrival* (DoA) of the signals are computed using one of the source location algorithms described in chapter 4. The unitary ESPRIT algorithm has been used in the computation of figure 5.17. It can be seen from the figure that the signals are spread over a range of angles with a cluster centred at $-8°$ and at a delay of 1.8 μs. The delay equates to a distance of $d = c \times t$ m, where $c = 3 \times 10^8$ m/s and $t = 1.8$ μs , thus $d = 540$ m. Knowing the angle and distance, scatterers can be located and identified using a map. In this case the scatterer is a multi-storey office block. The range of angles

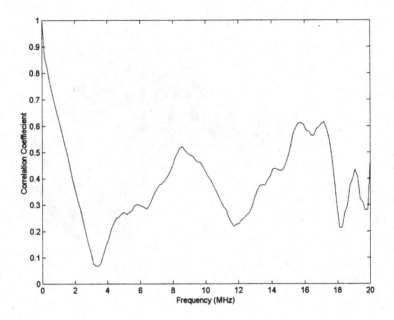

Fig. 5.16   Frequency correlation of a measured channel in the city of Bristol, UK.

the signal is spread over is quantified as *route mean squared angular spread*, σ_θ, and defined by the equation below.

$$\sigma_\theta = \sqrt{\frac{\sum_{k=1}^{K}(\theta_k - \psi_a)^2 \cdot \hat{p}(\theta_k)}{\sum_{k=1}^{K}\hat{p}(\theta_k)}} \qquad (5.18)$$

where

$$\psi_a = \frac{\sum_{k=1}^{K}\theta_k \cdot \hat{p}(\theta_k)}{\sum_{k=1}^{K}\hat{p}(\theta_k)} \qquad (5.19)$$

and $\hat{p}(\theta_k)$ is the power in the kth ray arriving at angle θ_k. The form of this equation is the same as that used for computing RMS delay spread in [100, 101] and is often considered as the angular domain equivalent.

The above impulse response and scatter map were computed over a 5 MHz bandwidth, however, the measurements where conducted over a 20 MHz bandwidth. With the availability of measured channel data over a 20 MHz bandwidth, the angles of arrival of the signals can be computed using 5 MHz bands at either

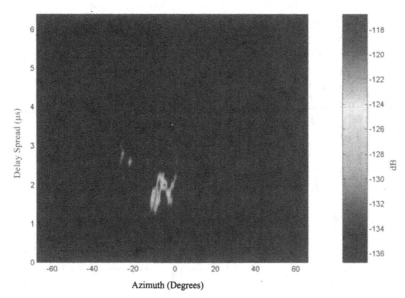

Fig. 5.17 Example scatter map of a 3G cellular system operating in an urban environment.

end of the 20 MHz bandwidth, i.e., separated by 15 MHz. An example of such a plot obtained from the above-mentioned measurement campaign is shown in figure 5.18. The two bands could be used as uplink and downlink transmission bands such as that in 3G UTRA. This operating mode is known as *frequency division duplex* (FDD).

The cause of this variation with frequency is two-fold. Firstly, the electrical characteristics of the array change with frequency which lead to the signals being subject to a frequency dependent distortion. Secondly, the radio channel characteristics are also frequency dependent, as can be observed in the channel frequency response in figure 5.15, which will lead to the received signal vector being frequency dependent and which is then reflected in the computed directions of arrival of the signals.

It is evident from the figure that different array weights are required for each link since the uplink and downlink directions of arrival differ. This presents a problem when designing smart antennas for such systems since, for the uplink, the channel characteristics can be extracted from the received signals because they have already propagated through the channel. This is not the case for the downlink and, because the channels are not the same, the parameters cannot be accurately extracted from the uplink. Without suitable compensation this

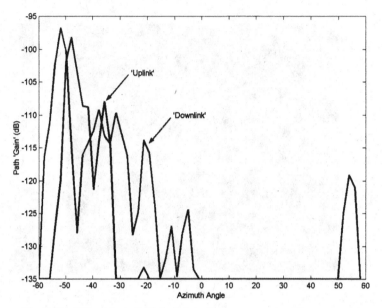

Fig. 5.18   Example uplink and downlink power azimuth spectrums.

will lead to suboptimal performance of the downlink beamformer. Techniques for compensating this are described later in this chapter.

5.4.4   Uplink-Downlink Channel Modelling for FDD Systems

It is possible to form a statistical model of the DoA-DoD offset that occurs between the uplink and downlink by using the channel measurements described in the previous section and in reference [99]. Such a model is useful in the simulation of smart antennas in FDD systems since it enables the downlink *directions of departure* (DoD - equivalent to the uplink's DoA) to be modelled from the uplink *power azimuth spectrum* (PAS) such that the offset of the angles approximates the histograms in figure 5.19. An application of this model is in the simulation of smart antenna systems applied to FDD air interfaces. It is proposed here that a Laplacian statistical distribution models the primary downlink DoD from the uplink PAS. It has also been shown in [28] that this distribution accurately models the distribution of DoA arriving at a base station employing an antenna array on the uplink. The Laplacian histogram takes the general form shown by

$$P(\theta) = \frac{k}{2 \cdot c} e^{-\frac{|\theta|}{c}} \tag{5.20}$$

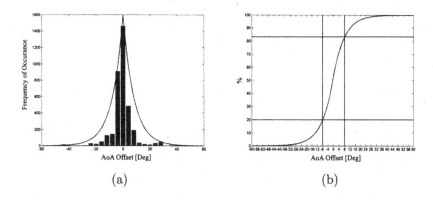

Fig. 5.19 Laplacian distribution for $c = 5$: (a) histogram; (b) CDF.

where c is a constant and the standard deviation, $\sigma = \sqrt{2} \cdot c$ [103] and k is a scaling factor.

Using the instantaneous measurement data previously described, which yields 3600 measurements, c has been empirically estimated to be 8 for an angular resolution of 8°. The angular resolution refers to the angular resolution that the source estimation algorithm can resolve into individual sources. This yields the histogram shown in figure 5.19, which has been plotted in conjunction with the angular offset between the primary DoA and DoD computed from measured data for comparison.

The measured data has been processed as follows. The uplink and downlink power azimuth spectrums are computed. The angles of the most powerful DoA and DoD are found and the difference between them computed. Setting $k = 500$, a good agreement between the two distributions can be seen. The *cumulative density function* (CDF) resulting from the synthetic distribution is shown in figure 5.19. The CDF enables the outage of the system to be computed. This is defined as the probability the DoA/DoD angular offset of the primary angular signal is larger than the 3 dB beamwidth. In this case an eight-element array has been used so the 3 dB beamwidth is approximately 16°. Thus, from the figure, which has been plotted using the Laplacian model, it can be seen that outage will occur for 35% of occasions. This corresponds with the outage compiled directly from the measured data [104]. Thus, the perturbations between the primary uplink and downlink angles are shown to be modelled by a Laplacian random process. The primary downlink DoA can therefore be modelled from the uplink by the equation below

$$\theta_{DL1} = \theta_{UL1} + \Delta\theta \qquad (5.21)$$

where θ_{DL1} and θ_{UL1} are the primary downlink and uplink DoAs respectively, and $\Delta\theta$ is the angular perturbation modelled as a Laplacian random process with $k = 500$ and $c = 8$. A similar process can be adopted to model the offset between the remaining secondary DoAs and DoDs.

5.5 TRANSMIT BEAMFORMING

In the previous section, the frequency dependence of spatial propagation has been demonstrated for a mobile communications system operating in an urban environment. It is a particular problem for FDD operation. Furthermore, the focus of the discussion in this book has been on receive beamforming. The following section discusses transmit beamforming and proposes solutions for the frequency dependency described above.

The goal of the downlink (transmit) beamformer is to send multiple co-channel signals from the base station, through the propagation environment, to the users so each user receives the desired *quality of service* (QoS) whilst minimising interference to neighbouring cells. This is in contrast to the uplink where the objective is to spatially filter each received signal to reduce the amount of co-channel interference from intra-cell and inter-cell users. This is achieved in both cases by adjusting the radiation pattern by means of applying complex weights to each antenna element. In the case of uplink beamforming, the weights are directly updated to the current user and interference situation since the received signals have propagated through the channel enabling channel estimation and hence optimum array weights to be computed. This is not possible on the downlink since the channel is generally unknown, as previously described.

Time division duplex (TDD) systems employ the same frequency for the uplink and downlink, thus the two channels can be assumed reciprocal if they are estimated within the channel coherence time (the time over which the channel is assumed to remain correlated). In the third-generation mobile communications system TDD specification, the duplex time is variable between 667 μs and 9.33 ms depending upon the channel symmetry [105] which falls within the coherence time of typically 200 ms for pico-cells [106]. *Frequency division duplex* (FDD) systems separate the uplink and downlink into two frequency bands which are often spaced apart by many megahertz. GSM employs a duplex separation of between 10 MHz and 85 MHz (depending upon the spectrum allocation) [107] and third-generation FDD mobile communications employ a spacing of typically 190 MHz [108]. Since the frequency spacing is much larger than the coherence bandwidth B_c, the instantaneous fading is significantly different therefore precluding the downlink channel from being accurately inferred by the uplink. It is possible to exploit the underlying time averaged azimuth distribution of signals arriving on the uplink as these are relatively stationary between the frequency bands.

Alternatively feedback can be applied on the uplink that reports the downlink channel characteristics.

As well as the de-correlation between the uplink and downlink discussed previously, the downlink beamforming problem is further complicated when required to support connectionless, packet-orientated services, i.e., discontinuous transmissions, as shown by Hugl and Bonek in reference [109]. The reference illustrates that the instantaneous angular distribution of users may differ between the uplink and downlink since a user's service requirements may be asymmetric as shown in figure 5.20. The reference shows that good downlink beamforming gain is achievable when priority is given to optimising the QoS of connection-orientated, symmetric traffic, i.e., speech, as the user distribution is known.

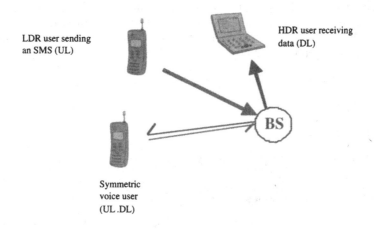

Fig. 5.20 Asymmetric user distribution in mixed traffic networks. LDR = low data rate, HDR = high data rate.

A common issue with most major air interfaces such as GSM and UTRA is the requirement to continuously supply pilot channel transmissions to all areas covered by a sector or cell for purposes of channel estimation and synchronisation. In the case of UTRA this includes the pilot, synchronisation and paging channels. Otherwise, subscribers in the coverage area may not be able to detect that a base station is available to provide a service. This is known as pilot pollution [110] and can severely impair network performance due to the additional interference it creates. In [111] the pilot was broadcast using an omni-coverage antenna making it available to the whole coverage area, but pilot power pollution will consequently be at a maximum.

A comprehensive review of downlink beamforming schemes for FDD air interfaces is given later in this section. It is shown that *blind* transmission beamformers give satisfactory performance when the uplink channel parameters are temporally averaged and that feedback based schemes operate well in channels with low Doppler spread, i.e., low mobility of the users and scatterers. Consequently, performance is compromised during short packet transmissions and highly mobile users.

The beamforming techniques introduced so far have focused on the uplink where the array is receiving the signal. To supply the downlink (transmit) capacity demanded by multimedia mobile networks, downlink (transmit) beamforming is also required. As mentioned previously, in the case of FDD systems (such as UTRA [112] and GSM [113]), the uplink and downlink are usually separated by a frequency offset outside the channel coherence bandwidth, which means that instantaneous fading is uncorrelated between the uplink and downlink. Therefore, the uplink beamformer weights will be sub-optimum downlink weights. Specialised beamforming techniques are therefore required for these systems. Figure 5.21 illustrates the classification of existing downlink beamformers for FDD systems according to the type of channel information and beamforming algorithm adopted. The diagram shows three primary techniques:

- feedback based;

- blind; and

- switched beam.

Each are reviewed in the following sections.

5.5.1   Blind Techniques

Blind downlink beamforming techniques compute the downlink weight vector without any knowledge of the downlink channel. The uplink channel parameters such as angle of arrival or temporally derived array weights are used to compute the downlink weights from the uplink channel parameters, as shown in figure 5.22. Three stages are shown. Firstly, the uplink channel parameters are estimated. This is performed by means of a temporal or spatial beamformer. Either the parameters are used directly to compute the uplink and then downlink weights (as in TDD systems), or they are subject to a frequency transformation process, before the downlink weights are computed.

The most basic technique is to simply use the complex conjugate of the uplink weights for the downlink as shown by equation (5.22)

$$W_D = W_U^* \qquad (5.22)$$

where $\mathbf{W}_D$ is the downlink weight vector, $\mathbf{W}_U$ is the uplink weight vector and $\cdot^*$ signifies the complex conjugate.

Zetterberg and Ottersen have used a spatial reference beamformer to generate the uplink weights in [114]. The simulation first computes the angular positions

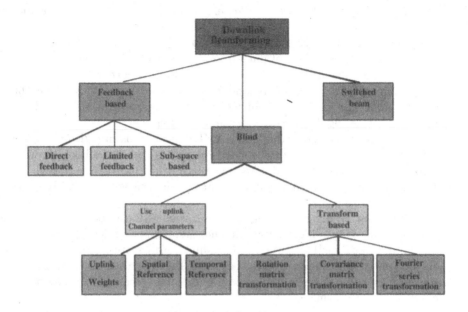

Fig. 5.21   Downlink beamformer classification.

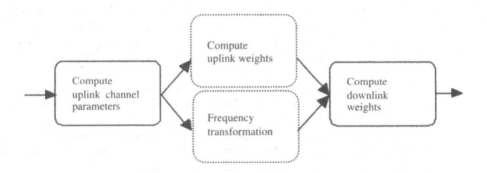

Fig. 5.22   Downlink beamforming concept.

of the arriving signals. Each user is then allocated a channel according to their angular positions. The downlink weights are then computed before transmission. Results show an increased spectrum efficiency by a factor of four when using a seven-element ULA array in an environment with 1.6 μs of RMS angular spread. It is also shown that in order to retain this spectral efficiency in environments of high angular spread, more antenna elements are required to provide additional degrees of freedom.

Lindmark, Ahlberg, Nilsson and Beckman have also investigated this technique [115] using measured channel data from a sub-urban environment and a spatial reference uplink beamformer. Results show downlink carrier-to-interference ratio improvements of between 6.7 dB and 29 dB. The authors indicate that this level of performance is likely to be unachievable in urban environments due to the increased scattering density (and hence angular spread).

Once again referring to the TSUNAMI II field trials in [27], uplink temporal reference derived weights were applied directly to the downlink. Here, a mean received power gain of 14.4 dB was measured in a sub-urban macrocell using an 8-element array compared to a single antenna element with a 3 dB beamwidth of 120°.

Thompson, Grant and Mulgrew have reported the performance trade-off between employing uplink spatial reference and temporal reference derived weights on the downlink in [116]. Here, the uplink weights are again used directly on the downlink. Simulation results show that the spatial reference algorithm gave favourable results, however, the amount of angular spread and the size of the FDD offset limited performance.

Angular spread causes performance impairments when it becomes much greater than the radiation pattern null width. Hugl, Laurila and Bonek have proposed a scheme in [13] to improve downlink beamformer performance by means of null broadening. Here, simulations have shown an SNIR gain of up to 8 dB is achievable by applying null broadening. This technique is, however, limited by the number of degrees of freedom available. An M-element array has M degrees of freedom, where one DoF is required to direct the main beam. Hence $M - 1$ degrees of freedom are available for null steering and broadening. Thus for $M = 8$, seven degrees of freedom are available for nulling co-channel users. Assuming one additional null is directed to co-channel users, three co-channel users can be supported in null broadening mode. This technique is therefore particularly suited to very low user densities or to null out hot spots.

Kitahara, Ogawa and Ohgane have proposed a novel beamforming technique in reference [117] that reduces the sidelobe level of the downlink radiation pattern using an adaptive signal cancellation process. This overcomes the DoF limitations associated with null broadening. A sidelobe level reduction of 15 dB compared to conventional uniform weighting is achieved using a six-element array, thus considerably reducing the interference received by many co-channel users. The main lobe is, however, broadened by 5°.

The discussion so far has not considered the effects of the array manifold that occurs between the uplink and downlink. Figure 5.23 shows the difference between the uplink and downlink radiation patterns for a uniformly weighted $\lambda_{ave}/2$ spaced eight-element array with an FDD spacing of 190 MHz. λ_{ave} is given by equation (5.23).

$$\lambda_{ave} = \frac{\lambda_{UL} + \lambda_{DL}}{2} \tag{5.23}$$

where λ_{UL} is the wavelength of uplink frequency, and λ_{DL} is the wavelength of downlink frequency. Perturbations in nulls are clearly seen and main-lobe perturbations become more prominent when steered off boresight. The frequency dependent array response that leads to the pointing errors shown in the figure has not been considered in the techniques reported so far. By applying a frequency transformation to the temporally averaged uplink spatial covariance matrix, pointing errors are reduced thus leading to an improved performance. Three principal methods of achieving this are now discussed.

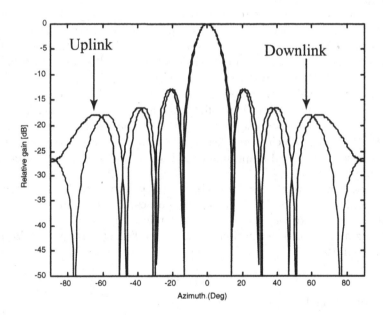

Fig. 5.23 Uplink and downlink radiation patterns with $d = \lambda_{ave}/2$.

Hugl, Laurila and Bonek have proposed a spatial covariance matrix transformation technique in [118] and [119]. Here the averaged uplink spatial covariance matrix is split into separate signal and noise matrices. The interference and user power azimuth spectrums are then estimated for each using the Capon DoA technique as described in chapter 3 and in [40]. The PASs are then shaped to reduce the beam-pointing error towards the user and improve null definition towards co-channel users. This is done by allocating the dominant DoA to the desired user and assuming other DoAs correspond to co-channel users (the loss of diversity gain induced by only using the primary DoA is discussed in the final section of this chapter). The user's downlink spatial covariance matrix is then computed using equation (5.24).

$$\mathbf{R_d} = \sum_{\theta} \hat{\rho}_{mod}(\theta)\mathbf{a_d}(\theta)\mathbf{a_d^H}(\theta) \tag{5.24}$$

where $\hat{\rho}_{mod}$ is the user's modified power azimuth spectrum, $\mathbf{a}_d$ is the downlink steering vector and $\cdot^H$ is the Hermitian transpose of a vector. The co-channel user downlink spatial covariance matrix, $\mathbf{R}_d$, is also estimated in the same way. The optimum downlink weights are then computed. Results in [118] show a 10 dB performance degradation when the spatial covariance matrix has no temporal averaging applied to it.

A second transformation technique expressing the uplink covariance matrix, $\mathbf{R}_u$, as a Fourier series expansion has been proposed by Goldberg and Fonollosa in [120]. Fourier coefficients of the spatial covariance matrix are estimated at the uplink frequency by means of least squares estimation. The coefficients are then used to compute the spatial covariance matrix at the downlink frequency, $\mathbf{R}_d$. Simulations show an average signal-to-noise-plus-interference ratio improvement of 20 dB above a single element for both low and high angular spread environments. Performance is, however, based upon the assumption that the signal phase is constant over the FDD spacing which is an idealistic assumption since it was shown earlier in this chapter that signal amplitude (and therefore phase) is heavily dependent on frequency.

A rotation matrix based transformation is proposed by Rayleigh, Diggavi, Jones and Paulraj in [121]. Here, the phases of the uplink spatial covariance matrix are rotated corresponding to a dominant DoA. This is accomplished by applying a linear transformation (Φ) to the uplink steering vector ($\mathbf{a}_u$) as shown in equation (5.25).

$$\mathbf{a_d} = \Phi\mathbf{a_u}(\theta) \tag{5.25}$$

where $\Phi = \mathrm{diag}(1 \; e^{j\phi_2} \; \ldots \; e^{j\phi_M})$

$$\phi_i = \left(\frac{1}{\lambda_d} + \frac{1}{\lambda_u}\right) \cdot \psi \tag{5.26}$$

and λ_u and λ_d are the uplink and downlink wavelengths respectively and

$$\psi = \frac{2\pi}{\lambda_u}d\sin\theta \tag{5.27}$$

It is assumed, however, that a dominant DoA exists and that a linear transformation is valid. Hence, in non line-of-sight scenarios and large FDD offsets, performance degrades.

The performance of these transformation techniques are compared in [122] where it is shown that the spatial covariance matrix transformation yields the best performance, particularly in non line-of-sight scenarios with more than one scattering centre.

Finally, Utschick and Nossek estimate the downlink steering vector by least squares estimation in [123] which is based on an algorithm proposed in [124]. This uses the difference between the uplink and downlink radiation patterns as the error criteria. First, the downlink beam pattern is computed at the uplink frequency using the power minimisation criteria discussed earlier. The beam pattern is then estimated for the downlink frequency by means of least squares estimation. The beam weights are finally computed and applied. Results show a marked improvement over direct application of uplink weights.

5.5.2 Feedback Based Techniques

The methods described above are not suitable for all propagation environments since large differences between the uplink and downlink channel parameters will yield a sub-optimal estimation of the downlink beam weights [104]. One technique that can be used to improve downlink beamforming is to employ a feedback signal from the mobile to base station to inform the base station of the downlink channel estimate. This has been proposed by Gerlach and Paulraj in references [125] and [126]. The operation is described as follows. Initially a base station will use weights directly from the uplink. If the interference at the mobile is too high it will signal to the home base station to enter probing mode so tracking of the mobile on the downlink is improved. The base station will then transmit using a number of test weights whilst the mobile sends its response on the uplink frequency back to the base station. Probing stops once the mobile has sent a satisfactory response.

A drawback with this method is the level of modification required to the network protocol (however, the UTRA FDD air interface has incorporated a downlink pilot channel allowing channel estimation to take place at 1.5 kbits/s feedback on the uplink [112]). Also the resources required to transmit the probing signals and associated response will mean a reduction in network capacity. Furthermore, a feedback rate of 20 kbits/s is required for a mobile moving at 42 km/h with an angular spread larger or equal to the beamwidth [126], therefore adding considerable overhead to the uplink. Gerlach and Paulraj have also proposed a method of reducing the feedback rate by a factor of around 100 (making it possible to deploy in UTRA networks) in reference [127] which exploits the slowly varying channel subspace estimate. The subspace method works by considering the average signal-to-interference ratio at the mobile. By maximising this parameter, the probability of it falling below some threshold when compared to the instantaneous SIR is much reduced therefore considerably reducing the feedback rate.

Choi, Perreau and Lee present a semi-blind downlink beamforming technique in reference [128] that overcomes the performance degradation caused by feedback delay (typically one or two slot periods depending upon the processing delay and channel synchronisation) in channels with high Doppler spread. Here, the feedback rate is equal to the slot rate and is assumed to be within the channel coherence time. A linear predictor is used to predict the downlink impulse response. This information is then used together with estimates of the uplink power azimuth spectrum to compute the downlink weights. Simulation results show a bit error rate improvement over the blind technique reported in [129] for Doppler spreads of <100 Hz. The blind technique yielded a bit error rate improvement factor of around 2 for higher Doppler spreads compared to the semi-blind scheme.

Auer, Thompson and Grant have compared two downlink beamforming techniques in an FDD-CDMA system by simulation in reference [130]. A three-tap delay line channel is assumed. The *single tap algorithm* (STA) transmits a training signal to the mobile on the three best bearings (as determined by the uplink channel estimate) in turn. The mobile selects the best bearing and feeds back the results to the base station (simplified feedback is therefore implemented here). This bearing is then used for future transmissions. The *multiple tap algorithm* (MTA) transmits on all three bearings simultaneously. The mobile then utilises a rake receiver to recombine the channel energy. Feedback is therefore not required.

Performance is degraded in all cases with increased angular spread. STA was found to give the best performance because the signal energy was split between fewer taps therefore undergoing less angular spread. It, however, has the inherent disadvantage of using feedback, namely increased network overhead.

5.5.3 Switched Beam Techniques

By way of an alternative to the computationally intensive adaptive downlink beamforming approaches discussed so far, a switched beam system is considered here. The system architecture is identical to that of an uplink switched beam system (see chapter 4). Tirrola and Ylitalo have appraised a system using two beams over a 120° sector based on a two-port Butler matrix [131]. Here, the downlink beam is selected as the uplink beam that gives the highest received signal strength. Results show a performance improvement for channels where the angular spread is considerably larger than the beamwidth. The authors also conclude that two beams per sector is not sufficient as cusping loss impairs service to users located in this region.

The key advantages of these systems over the adaptive techniques presented in previous sections are:

- *Inherent numerical stability.* Systems do not require signal processing since beam selection can be based on RF, analogue and control logic. Consequently, algorithm robustness is not a design issue.

- *Considerable reduction in computational overhead and delay.* As stated above, complex signal processing algorithms are not required.

- *Simple implementation.* These systems can be retrofitted to existing base stations without the complexity associated with digital or adaptive beamforming schemes.

Furthermore, a minimum isolation between co-channel users of around 13 dB is achievable (indicated by the sidelobe levels in figure 4.6) whereas fully adaptive techniques can suffer from poor null depth or high sidelobes due to angular spread and non-ideal array calibration. Hence, there is no degrees of freedom limitation in switched beam systems when subject to high user densities.

The simulation results reported in [32] have shown an area outage improvement of more than 10 dB above an omni-sector antenna when a four-beam, 30° beamwidth system is deployed in a macro-cellular environment. Mootri, Stutzle and Paulraj report a capacity gain of four to five times that of a standard sectorised IS-95 forward link in reference [132] when a twelve-beam system is employed. The field trials reported in [31] show a 3 dB improvement in downlink carrier-to-interference ratio for a similar system. Furthermore, a *grid of beams* system was implemented in the TSUNAMI II field trials system [27]. Here, a mean received power gain of 14.7 dB above a single element system was recorded.

The application of switched beam based downlink beamforming in a 3G UTRA system forms the basis of a beamforming example in chapter 6.

5.5.4 Downlink Signal Distribution Schemes

The previous section focused on downlink weight estimation algorithms where weight optimisation criterion has not been discussed, i.e., strategies of translating uplink information to downlink have been discussed but the computation of the downlink beamforming weights from this input forms the discussion of this section. In this section the merits of the two principal downlink weight optimisation criteria are considered:

- transmit maximum energy towards the user; and

- minimise energy towards co-channel users whilst simultaneously transmitting enough energy towards the user.

The objective of the first criterion is to maximise the user's SNR as defined by equation (5.28) [129].

$$SNR_{max} = max \frac{p\mathbf{W_d^H R_u W_d}}{\sigma^2} \tag{5.28}$$

where p is the transmit power, $\mathbf{W}_d$ the downlink weight vector, $\mathbf{R_u}$ the uplink spatial covariance matrix and σ^2 is the noise power. This occurs when $\mathbf{W_d^H} = \mathbf{e_1^H}$ where $\mathbf{e}_1$ is the principal eigenvector of $\mathbf{R}_d$ [133] and, assuming reciprocal uplink and downlink channels, $\mathbf{R}_d = \mathbf{R}_u$. This technique, however, results in a loss of diversity gain at the mobile because only the channel tap corresponding to the maximum eigenvalue is used in the computation. In practice some diversity gain does occur due to FDD channel de-correlation and angular spreading. Thompson, Hudson, Grant and Mulgrew have proposed a technique of improving the diversity loss in reference [129]. Here, equal power is transmitted in the corresponding directions of each channel tap. This results in a sub-optimal user SNR but an improved diversity gain. In the case of no angular spread, the proposed beamformer provides a 3.5 dB improvement over the maximum SNR solution. A 2 to 2.5 dB improvement is shown with 30° of angular spread.

The second algorithm constitutes an optimisation problem where all users require a minimum average SNIR. This is interpreted as minimising the total power transmitted by each base station whilst providing the required SNIR to the users. This is formally defined by equation (5.29) [37].

$$min \sum_{i=1}^{I} \|\mathbf{W_i}\|^2 \qquad (5.29)$$

such that

$$\frac{\mathbf{W}_i^* \mathbf{R}_{i,k(i)}}{\sum \mathbf{W}_n^* \mathbf{R}_{i,k(i)} \mathbf{W}_n + \sigma^2} \geq SNIR_{TH}, \qquad i = 1, \ldots, d \qquad (5.30)$$

where $\mathbf{W}_i$ is the ith user's downlink weight vector, $\mathbf{R}_{i,k(i)}$ is the uplink spatial covariance matrix of the kth user, W_n is the weight vector for the nth co-channel user and $SNIR_{TH}$ the SNIR threshold of the ith user.

Rashid-Farrokhi, Lui and Tassiulas have presented a solution to the above optimisation problem in [134] which relates to algorithms used for iterative power control in interference limited cellular systems. Bengtsson and Ottersen have reformulated it as a *semi-definite optimisation problem* in [37]. The two solutions are compared in [135] where it is reported that the first solution converges the fastest in simple scenarios, but convergence time rapidly increases with complexity. The second strategy was found to have several advantages:

- computational complexity is fairly independent of constraints; and

- possible to consider more constraints such as power on each antenna element.

The main drawback of implementing this optimisation solution is the requirement for all channel parameters to be known and therefore collected at a central server where the solution is computed and then distributed to the basestations. This adds considerable network overhead and computational delay due to the amount of data that is required to be monitored and the computational complexity of the numerical solution.

Two signal distribution schemes have been compared in [136] in a simulated UTRA FDD network with a 120° sector, eight-element uniform linear array with

$\lambda/2$ inter-element spacing employed at the base station. Results show the sub-optimal scheme to improve E_b/I_o by 5 dB over a single element. The optimal scheme provided a further 5 dB improvement. However, the sub-optimal scheme outperformed the optimal one in low E_b/I_o scenarios due to poor spatial signature estimation associated with the algorithm used in the later scheme.

5.6 CHAPTER SUMMARY

This chapter has discussed practical considerations when implementing array systems. The discussion has used a mobile communications as a basis of the discussion. The main topics covered have been:

- signal processing constraints;

- implementation constraints, which has included system linearity, calibration, mutual coupling and circular arrays;

- radiowave propagation, where a channel model was developed that included multiple antennas and wideband channel effects that impact beamformer design for FDD wireless systems; and

- transmit beamforming, where several techniques have been introduced that attempt to reduce the impact of radiowave propagation over frequency.

This has set the scene for the following chapter where the application of antenna arrays is considered.

5.7 PROBLEMS

Problem 1. Given a 35 m mast with an equipment room at its base, calculate the increase in the noise figure of the BS receiver if either an LDF4-50A or LDF5-50A feeder were used (allow 5 m of extra feeder to terminate in the equipment room, assume no active components are used at the top of the mast).

Problem 2. Calculate the reduction in noise figure by installing an LNA near the antenna array (allow 0.1 dB feeder loss between the antenna and the LNA input and an LNA noise figure of 0.5 dB with gain of 30 dB) for the parameters in Problem 1.

Problem 3. State the difference between static and dynamic calibration.

Problem 4. List the techniques that can be utilised to design a digital beamformer with calibration in mind.

Problem 5. Derive the Davies transformation (equation (5.9)).

Problem 6. What is meant by *angular spread*? Draw a diagram to illustrate how it is caused.

Problem 7. Describe the basic process of blind downlink beamforming.

Problem 8. List the key stages in feedback based downlink beamforming.

Problem 9. List the primary advantages and disadvantages of blind downlink beamforming algorithms.

6

Applications

6.1 INTRODUCTION

The application of antenna arrays to biomedical, communications, radar and sonar have been explored by many organisations worldwide and is well understood. Consequently, demonstrator systems have been developed and deployed and the results reported in the literature. The system configurations differ considerably in the choice of operating environments and applications, antenna and hardware architecture and algorithms which makes comparison between the results difficult, however, broad observations can still be made.

This chapter describes some practical examples of the application of antenna arrays to biomedical, radar, sonar and terrestrial and space communications applications. Each of the examples draws on many of the principles introduced throughout this book and aims to put into context the application of antenna arrays and related signal processing techniques for each of the fields of interest.

6.2 ANTENNA ARRAYS FOR RADAR APPLICATIONS

Radar was first conceived just before the beginning of the second world war. At the time it used frequencies in the VHF part of the radio spectrum (often in the 100 MHz to 200 MHz band). Consequently the physical dimensions of the antennas were tens of centimetres. As technology advanced, microwave operation became possible and many radars were assigned to the 3 GHz band termed

Adaptive Array Systems   B. Allen and M. Ghavami
© 2005 John Wiley & Sons, Ltd   ISBN 0-470-86189-4

S band. This meant that high gain antennas could be deployed with dimensions comparable to the lower gain antennas operating in the VHF band. The higher gain antennas meant that a narrower beam width was achievable and the added path loss introduced at these higher frequencies could be overcome. Consequently, by rotating the antenna towards potential targets, the angle of targets could be resolved to within the beam width of the antenna. Rotation was achieved by mounting the antenna on top of an electric motor. Such radars are akin to those often seen on boats and at airports. The antenna could also be rotated in elevation enabling a two-dimensional 'fix' of the target to be determined. Signal processing techniques would then enable the range to be computed.

More recently, rotating antennas have been replaced with antenna arrays, thus eliminating the need for mechanical components. Furthermore, a planar or spherical array enables a target to be located in both azimuth and elevation. A planar array forms part of the 'Patriot' radar missile defence system used by the US military. This consists of a planar antenna array with some 25000 elements. A planar array does not enable targets outside of a beam width of 120° to be detected. This can be overcome by mounting several arrays to form a pyramid or to use a spherical array.

Electronically scanning the beam in azimuth and elevation enables targets to be initially detected and identified. The system can then switch into tracking mode to track the movement of the target. This information then enables strategic decisions to be made. By employing multiple beamformers, multiple targets can be detected and tracked simultaneously. The Patriot's beamforming architecture enables 100 targets to be simultaneously tracked.

6.3 ANTENNA ARRAYS FOR SONAR APPLICATIONS

So far much of the discussion has been on antenna arrays for wireless applications and electromagnetic signal propagation has been assumed. Sonar has much in common with radar, except that it operates in water and uses acoustic (sonic) signals. Thus, instead of antennas, loudspeakers (projectors) and microphones (hydrophones) are used and collectively referred to as *transducers*. Also, the propagation of the signals differs although the basic rules of reflection, refraction, diffraction and attenuation still apply. The propagation characteristics of the signals are complicated by the variations that occur with depth and impact of sea water and the oceans surface and bed. The velocity of sound in the ocean is in the region of 1500 m/s, but the exact velocity depends upon the pressure, temperature and salinity of the water. Pressure and temperature effects the *bulk modulus*, B (N/m^2), and the density, ρ (kg/m^3) of the water, whereas salinity effects the density. The velocity, v, is given by

$$v = \sqrt{\frac{B}{\rho}} \quad \text{m/s} \tag{6.1}$$

Note that the wavelength of acoustic signals propagating in the ocean is in the range of 7.5 cm to 150 m which is equivalent to a wireless array operating over the frequency range of 2 MHz to 4 GHz and presents challenging parameters for sonar systems design.

Sonar systems, like radar, frequently transmit pulses to probe the environment, and are of short duration and have a corresponding bandwidth that covers much of the acoustic spectrum. Consequently the wideband array processing techniques described in chapter 3 can be applied to sonar systems.

Applications of antenna arrays to sonar systems include:

- *Fishing.* Locating and identifying shoals of fish.

- *Marine Exploration.* Locating wrecks and other hidden objects under the ocean.

- *Submarines.* Guidance through complex terrain and defence operations such as target location, identification and tracking.

- *Seismic exploration.* This is a land-based application where sonar is used in earthquake monitoring and geological exploration.

Thus, the array may be mounted underneath the vessel or trailed behind the vessel or submarine. It is important that the sensors are isolated from vibrations caused by the vessel's equipment as this is a source of acoustic interference. Therefore, an acoustic isolation section must be positioned between the vessel and the sensors, an example of which is shown in figure 6.1.

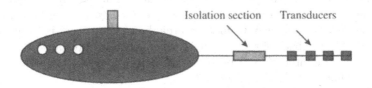

Fig. 6.1 Sonar array and isolation section for submarine applications.

6.4 ANTENNA ARRAYS FOR BIOMEDICAL APPLICATIONS

Antenna arrays, together with beamforming, have unique capabilities that enhance the performance of military and commercial communication systems such as remote sensing, radar systems, and recently, innovative biomedical applications.

In this section two applications of antenna array systems in biomedical engineering are discussed. Both techniques are used for medical imaging. However, they employ different frequencies, and consequently, different array types.

6.4.1 Medical Ultrasonic Arrays

The basics of ultrasonic transducer array design in the frequency range useful for medical imaging, 1 to 10 MHz, are explained in this subsection, and the interested reader is referred to reference [137] for more information.

6.4.1.1 Basic principles. An antenna array and beamformer allow broadband ultrasonic pulses to be generated and swept (scanned) through a body for medical imaging. The design of the array is dependent on the particular clinical application or target organ to be imaged. Image resolution increases with frequency, however, attenuation of energy by body tissue at the rate of 0.5 dB to 0.7 dB/cm/MHz dictates a compromise in final choice of the operating frequency.

The basic structure of an ultrasonic array is illustrated in figure 6.2. The figure illustrates the cross-section of the three major elements, with each consisting of up of two sub-diced elements, to maintain desirable aspect ratios. The basic principles for constructing a transducer to generate a short pressure pulse can be understood by considering the simple diagram of figure 6.3 illustrating four basic pressure pulses generated when a piezoelectric ceramic material is shock excited by an impulse voltage of approximately 100 V.

At each of the major surfaces of the ceramic, two pressure waves are generated travelling in opposite directions: a positive pressure wave and a negative pressure wave. If the ceramic is surrounded by air or water, most of the wave amplitude will be reflected at the boundaries of the ceramic, causing the pulse to ring for a long time.

The fraction of the pulse amplitude reflected at each interface is proportional to the difference in acoustic impedance between two materials on either side of the boundary. One way to make a broadband transducer is to attach a backing material to the back of the ceramic with an acoustic impedance close to that of the ceramic. The internal reverberations of the ceramic are then eliminated, allowing a very short pulse to be generated.

The acoustic impedance is defined as *the product of the density of the material and the longitudinal sound velocity. Non-destructive testing transducers* (NDT) generate very short pulses by this principle of attaching a very dense backing material. The backing material chosen should ideally not only closely match the

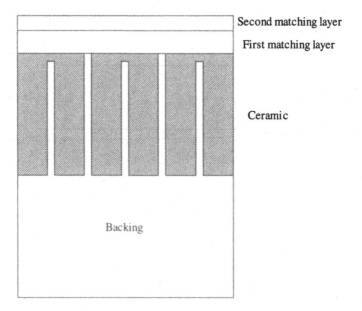

Fig. 6.2 Sub-diced linear array used in medical ultrasonic applications.

ceramic acoustic impedance but also be very absorptive so that no energy is left to be reflected from the non-ceramic end of the backing and create an artefact echo.

Although adequate for NDT, this ceramic construction does not have enough sensitivity for medical applications, since so much of the energy applied in creating the pulse is lost into the backing. The acoustic intensity in a material may be expressed by the following formula

$$I = \frac{1}{2}Zu^2$$
$$= \frac{1}{2}\frac{P^2}{Z} \tag{6.2}$$

where I is the intensity in watts/cm^2, Z is the acoustic impedance, u is the particle velocity, and P is the pressure. To increase the energy transmitted into the front medium it is necessary to make the tissue in front of the ceramic (the effective load) of a high impedance. An analogous situation is an electrical circuit, where maximum power is transferred from a source to a load when the impedance of the load equals that of the source. A technique to achieve this is by using an impedance transformer. In acoustics, the concept of acoustic impedance is used and transmission line sections become matching layers of intermediate impedance values. Thus, for medical ultrasonic applications, an optimum combination of

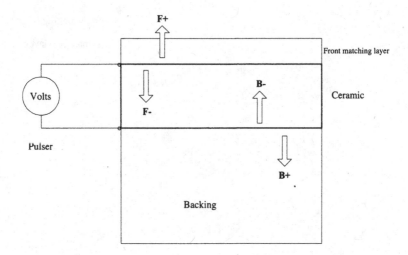

Fig. 6.3 Four basic pressure waves generated by a voltage impulse. '+' denotes compression wave and '−' stands for rarefaction wave.

broad bandwidth, short pulses, and good sensitivity can be achieved by using a light-weight impedance backing with two quarter wave matching layers on the front of the piezoelectric element.

One measure of the sensitivity of a transducer is *insertion loss*. Insertion loss is defined as the ratio of the power delivered to that part of the system following the transducer, before insertion of the transducer, to the power delivered to that same part of the system after insertion of the transducer. The term *transducer loss* is defined as the ratio of the available power of the source to the power that the transducer delivers to the load.

To measure insertion loss one must account for beam spreading diffraction losses, electrical impedance mismatches, etc. For medical ultrasonic transducers, insertion loss numbers fall in the range of 6 dB to 10 dB. More practically, when testing the individual elements of a multi-element array, one sets the target a fixed distance that is close to the elevation axis focal point, and defines a more empirical term called *loop sensitivity*. This is defined as the ratio of the peak to peak pulser voltage to the received (two way) pulse echo voltage peak to peak amplitude. Values for loop sensitivity are typically −25 to −45 dB.

6.4.1.2 Design of acoustic stack. The frequency output is related to the thickness of the ceramic material from which the transducer is constructed. However, if one attaches leads to the electrodes of a piece of ceramic in air and measures the electrical impedance as a function of frequency, one observes two resonances. The resonance at which the impedance reaches a minimum is called the *resonance frequency*, and the higher frequency resonance where the impedance reaches a peak is called the *antiresonance frequency*.

The position and shape of the resonance frequency depend dramatically on the length, width, and thickness of the ceramic slab. The antiresonance (or parallel resonance) frequency is fundamentally related to the ceramic thickness, and not influenced by modal coupling.

For a short impulse response and broad bandwidth, ultrasonic transducers are designed with one or two matching layers on the face. The acoustic impedances of these layers are usually chosen based on transmission line theory.

6.4.1.3 Materials.

The acoustic materials used to fabricate a transducer's front and back layers are typically made from polymer resins. These are heavily loaded with powders to achieve the desired acoustic properties of acoustic impedance, velocity, density, and attenuation.

Broadband transducer designs are usually made with double matching layers on the front. The first layer may be made from epoxy loaded with tungsten or aluminium oxide. Acoustic impedance is an especially important quality of transducer material that serves as a measure of the pressure required per unit particle velocity, and thus of the stiffness or resistance to passage of pressure waves. The second matching layer materials are usually unloaded epoxy or urethane.

6.4.1.4 Thermal management.

Another practical issue in good transducer design is to avoid overheating the face of the probe which, may cause patient discomfort. The temperature at the face of the probe must not exceed 41°C. Phased array probes used for cardiology scanning tend to overheat more than linear arrays, since the same elements are used to generate every beam firing. Good thermal management techniques include inserting low profile conductive heat pipes or fins in the backing, and loading the backing and front layer polymers with powders of excellent thermal conductivity, such as aluminium nitride, or industrial grade diamonds.

6.4.1.5 Elevation focusing.

Focusing of the short axis, elevation dimension, determines the effective slice thickness of the image plane. As illustrated in figure 6.4, there are two approaches for focusing beam elevation along the short axis:

- internal focusing by the shape of the ceramic; or

- external focusing with a *room temperature vulcanising* (RTV) silicone lens.

An external convex lens made with material of a lower velocity than water is used with flat ceramic geometry in the short axis. These materials are in the RTV silicone family.

Alternatively, for internally focused array designs, with cylindrical shaped ceramic, it is desirable to have a neutral density face filler with impedance close

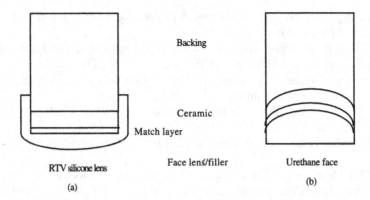

Fig. 6.4 Two alternate geometries to focus short axis elevation: (a) external focus; (b) internal focus.

to tissue or water. These requirements are met by certain families of castable urethane organic polymers. Some moderate level of acoustic absorption is actually desirable in the face material, because it helps attenuate multi-path reverberation from the tissue, which would otherwise create near field clutter in the image.

6.4.1.6 System considerations. There must be tight coupling between system and transducer design to achieve the best image quality in clinical applications. One of the most important design parameters in an ultrasonic array is the major element pitch, i.e., the distance between the major elements. A guideline to make a first estimate of a required element pitch is to decide how many channels are going to be used in beamforming.

Figure 6.5 shows an example of the angular sensitivity pattern for a 3.5 MHz phased array element. Comparison is made between air maintained in the kerfs (cut in material), versus the response if RTV material is allowed to wick into the kerfs, or alternately epoxy resin as a kerf filler.

Experience has shown that a good balance between resolution requirements and keeping down unwanted clutter acceptance is given by expanding the beamformer aperture at a rate allowed by the element angle roll-off at the −12 dB (two-way send/receive) point. It can be shown that having a larger number of beamformer channels is only useful at greater depths.

6.4.1.7 Tuning and electrical impedance matching. Another important aspect is electrical impedance matching. Since a multi-element array may have 100 to 200 elements, the individual elements are very small and subsequently have high electrical impedances (100 to 600 Ω). The array elements should be matched to

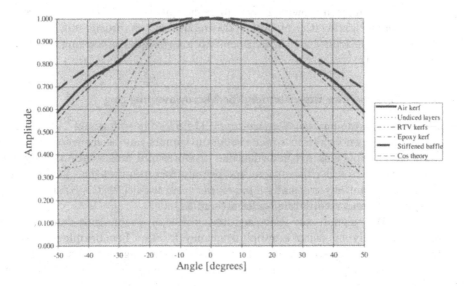

Fig. 6.5 Plot of angular sensitivity of individual array element and influence of boundary conditions and different kerf fillers.

the pulser during transmit, and the pre-amp impedance on receive for efficient energy transfer. For maximum energy transfer between the transducer and pulser or receiver, the load impedance of the transducer should be the complex conjugate of the electrical circuit coupled to it.

6.4.1.8 Summary. Image quality is a multi-parameter quantity, which depends on both the array transducer and the system performance. Each should be designed and optimised with the other in mind. Advances in array technology and system electronics are making possible ultrasonic real-time imaging with significantly improved image quality. Image quality is ultimately determined and limited by transducer performance.

The fundamental parameters of image quality are

- axial and lateral resolution;

- frequency content; and

- signal processing.

Clutter in beam sidelobes and grating lobes can limit dynamic range and contrast resolution capability. Careful matching of the transducer array element electrical impedances can optimise sensitivities and ultimate band widths achievable, so that 80 to 90% bandwidths are realisable.

Recent progress in growing larger single crystals with higher coupling constants should make bandwidths of over 100% realisable. A broadband, well-behaved response in the transducer array opens up many signal processing opportunities in the system.

6.4.2   Space-Time Beamforming for Microwave Imaging

Microwave imaging with space-time (MIST) beamforming for detecting backscattered energy from small malignant tumors has been investigated in the literature [138]. This method identifies the presence and location of significant microwave scatterers. The space-time beamformer assumes that each antenna in an array sequentially transmits a low-power *ultra wideband* (UWB) signal into the human body and records the backscattered signals.

The UWB signal may be generated physically as a time-domain impulse or synthetically by using a swept frequency input. The beamformer spatially focuses the backscattered signals to discriminate against clutter caused by the *heterogeneity* (i.e., non-uniformity) of the normal tissue and noise while compensating for frequency-dependent propagation effects.

The space-time beamformer achieves spatial focusing by first, time shifting the received signals to align the returns from a *hypothesised scatterer* (i.e., virtual scatterer) at a candidate location. The time-aligned signals are passed through a bank of finite-impulse-response (FIR) filters, one in each antenna channel, and summed to produce the beamformer output, similar to the process described in chapter 3.

The weights in the FIR filters are designed using a least squares technique so the beamformer passes the components of the backscattered signal originating from the candidate location with unit gain while compensating for frequency-dependent propagation effects.

The beamformer output is time gated to the time interval which would contain the backscattered signal from the candidate location, and then the energy is calculated. If a scattering object, such as a tumor, exists at the candidate location, a relatively large energy results. The beamformer is scanned to different locations by appropriately changing the time shifts, gating, and FIR filter weights. A display of energy as a function of location provides an image of backscattered signal strength.

In contrast to direct matched filtering techniques, MIST beamforming does not assume a specific backscattered signal. As a result the beamformer maintains robustness with respect to the uncertainty in the shape of the signal that arises from variations in tumor properties, such as size and shape.

The beamformer also provides an estimate of the backscattered signal from a specific location that may be further analysed to gain insight into local tissue properties. Such analysis, e.g., spectral analysis of the backscattered signal, is likely to be useful in tumor characterisation.

Two types of active microwave imaging techniques have been previously proposed for the case of breast cancer detection: microwave tomography and UWB radar techniques.

6.4.2.1 *Microwave tomography.* The goal of microwave tomography is the recovery of the dielectric property profile of the breast from measurements of narrowband microwave signals transmitted through the breast.   Whilst promising initial results have been reported, the solution of this nonlinear *inverse-scattering problem* requires challenging and computationally intensive reconstruction algorithms. Also, tomographic techniques generally require a large number of transmitting and receiving antennas to be distributed around the object of interest, which complicates the imaging of smaller breast volumes and breast tissue near the chest wall or in the upper outer quadrant of the breast where nearly half of all lesions occur.

6.4.2.2 *UWB radar techniques.* In contrast to tomography, UWB radar techniques do not attempt to reconstruct the dielectric-properties profile, but instead seek to identify the presence and location of significant scatterers in the breast.   MIST beamforming, which falls under this class of techniques, offers significant improvement over the previously reported techniques of simple time-shifting and summing of backscattered signals to create a synthetically focused signal.

Typically, a system sends a microwave pulse from a transmitter and receives it in the same location with the same antenna. Ground-penetrating radar for mine detection accomplishes detection with an antenna located typically just above the ground. The same basic principles used in ground-penetrating radar are also used in medical imaging.

6.5   ANTENNA ARRAYS FOR WIRELESS COMMUNICATIONS

The application of antenna arrays to wireless communication systems has formed the centre of much research over the past few decades. Arrays can either be placed at the transmitter, receiver, or both, however, there are considerable challenges associated with performing beamforming on handsets such as:

- cost;

- complexity; and

- performance.

High performance is difficult to obtain due to the large angular spread experience around the user which will limit the amount of spatial filtering that can be achieved. Also, the orientation and proximity to objects such as the user's body and handset will impact the performance of the array since it will distort the element's input impedance and radiation pattern and cause coupling between elements. Spatial diversity techniques, where the array is not used for beamforming but in a diversity transmission/reception mode as illustrated in figure 6.6 forms a very attractive alternative. The figure shows a transmitter with two antennas and each antenna transmits the signal. This gives the receiver two independent opportunities of receiving the signal, or receiving both signals and combining them such that the total signal provides an improved performance compared to either of the single signals received independently. Space-time coding techniques enable the transmitted signals to be sent simultaneously without interfering with each other. Optimal performance of these diversity based techniques relies on rich multipath associated with large angular spreads. Discussion on these topics is beyond the scope of this book but the interested reader is referred to reference [106] for further details. Within the context of mobile wireless networks, smart antennas can be

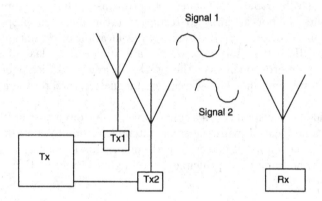

Fig. 6.6 Spatial transmit diversity concept with two transmit and one receive antenna. The system could employ additionally multiple receive antennas.

used to provide:

- increased network capacity;

- range extension;

- reduced susceptibility to multipath effects; and

- location awareness.

Network capacity is increased by controlling the interference level received from other users and base stations. In this case, smart antennas are used as a spatial filter and will therefore reduce interference arriving in certain directions. This operating mode is referred to as *spatial filtering interference rejection* (SFIR). A related operating mode is referred to as *space division multiple access* (SDMA), where smart antennas are used to separate signals, allowing different subscribers to share the same spectral resource, provided their signals are spatially separable at the base station. This mode of operation allows multiple users to operate on the same time slot (t_1), frequency (f_1) and, in the case of CDMA systems, the same code (c_1) as illustrated in figure 6.7. This allows any of these resources to be reused between beams. Note, however, that SDMA is unlikely to be a candidate for deployment in 3G mobile systems due to angular spread (described in chapter 5), which inhibits the attainable spatial isolation and can lead to a breakdown in SDMA mode, i.e., an undesirable level of interference between shared resources. The two examples in the following sections illustrate the application of SFIR to mobile communications networks which, unlike SDMA, has been shown to provide robust operation in mobile wireless systems. The first example considers uplink beamforming for second-generation mobile wireless systems and the second example considers downlink beamforming for third-generation mobile wireless systems. Smart antennas can also be used to provide range extension through the

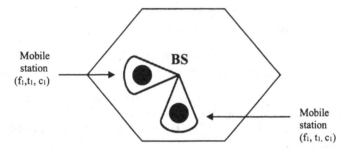

Fig. 6.7 SDMA principle.

inherent directional gain obtained from using an array. This is useful for rural cells which are required to cover larger areas than those in urban environments. The additional directional gain, D, provided by an M element array is approximated by equation (6.3).

$$D \approx 10 \log_{10} M \qquad (6.3)$$

Like SFIR, reduced susceptibility to multipath effects can be achieved by exploiting spatial filtering properties. In this case spatial filtering reduces the channel's delay spread. This is because signals arriving at angles outside of the main beam will be attenuated. These signals will have longer path lengths and would normally contribute to the longer delays of the channel impulse response. The consequence of reducing delay spread is that equalisation techniques are no longer required therefore simplifying the receiver design. Such an application has been considered for wireless local area networks as opposed to mobile communications networks [139].

An example presented later in this chapter considers location awareness in a mobile communications system. A location aware wireless network has the ability to locate and track network users and, due to recent legislation in the US, is a legal requirement for mobile communications networks. The location information is useful for location people who are making emergency calls. It can also be used to provide location-based services to the user such as informing the user of restaurants and shops within the vicinity. It is also useful for allocating network resources to busy areas (*hot spots*). By using the inherent directional information, smart antennas can assist in providing the location information.

6.5.1   Uplink Beamforming for Second-Generation Mobile Wireless Networks

As mentioned previously, beamforming offers several benefits to mobile communication systems:

- *Range extension.* This uses the array gain to increase the coverage area and is especially useful in rural areas.

- *Spatial filtering interference reduction.* This allows co-channel cells to be moved closer together, therefore reducing the frequency reuse spacing.

- *Spatial division multiple access.* This mode of operation employs several beams simultaneously, where each cluster of co-channel users is served by a beam, therefore separated by the amount of isolation offered between beams. This enables network resources to be shared between beams.

- *User location.* This allows users to be located and therefore has applications relating to emergency services, advertising and network resource allocation.

For such systems, an antenna array replaces a traditional omni-directional antenna at the base station. This may be a circular array if 360° continuous coverage is required, or a linear array for 60° or 120° sectored coverage. If the antennas are located above the average roof-top height, this is termed a *macro-cellular environment*. Alternatively with the antennas located well below the surrounding roof-tops, a *micro-cell* is formed. The main differences are the coverage area and propagation characteristics. Both configurations were considered during one of the largest trials of adaptive antennas in a communications system to date, a project funded by the European Commission termed TSUNAMI II (*Technology in Smart antennas for the UNiversAl Mobile Infrastructure II*) [139, 111], which is described here and the discussion is based upon reference [111]. A ten-element linear array was used with the elements mounted at spacings of $\lambda/2$. The eight central elements were active and the two outer elements were dummy elements to reduced the effect of mutual coupling. In order to increase the gain of each element, they were connected to seven other identical vertically mounted elements, thereby creating a directional beam in the vertical plane. The array was mounted on the mast with down-tilt to ensure the radiation pattern focused on the desired coverage area below. A low noise amplifier and duplexer arrangement were incorporated within the array.

Cables connect the array connected to the remaining RF and baseband equipment. A digital beamformer was employed at baseband, enabling the following algorithms to be tested:

- *Maximal ratio combining* (MRC). This algorithm performs maximum ratio combining on the uplink and uses the same weight vector for the downlink. It maximises the signal-to-noise ratio (SNR) at the combiner output, assuming that no interfering signals are present, by co-phasing the array signals and weighting them according to their SNR. In this implementation the weight vector is updated every four frames (18.46 ms).

- *Spatial reference beamforming using MUSIC* (MUSIC). Here MUSIC [41] is employed as a *direction finding* (DF) method in order to detect the number of signals present, and also estimate their *direction of arrival* (DoA). This DoA information is then used to synthesise a beam steered at the wanted signal and nulls in the direction of the other signals (usually interference). This beam synthesis is implemented separately for the uplink and downlink to account for the frequency difference between the uplink and downlink. In the TSUNAMI II test-bed, spatial smoothing, in the form of a Kalman filter, was also applied to the output of the MUSIC algorithm in order to enhance the tracking process.

- *Grid of beams* (GOB). Here thirteen fixed beams ($10°$ intervals from $-60°$ to $60°$ in azimuth) are synthesised by the beamformer and the beam with the highest output power is selected for reception. The same look direction is used for the transmit case.

- *Optimum Combining* (OPT). Here the DCS-1800 training codes are used as a reference signal to compute the Wiener optimum weight vector and the weights are adjusted to minimise the power of an error signal. The error signal is derived as the difference between the beamformer output and the local reference signal (training code). This is also known as temporal reference beamforming.

- *Temporal reference beamforming grid of beams* (AUC). This algorithm was developed by the University of Aalborg for use in the field trial system. In the AUC algorithm, 22 beams were formed over $60°$ of the useable beamwidth of the array, with beam selection based on wanted user training sequence correlation after 8 ms of averaging.

A test vehicle drove around a defined route while making a call on a GSM test set, i.e., this was based on second-generation mobile telephone technology. Interfering signals were also transmitted to enable the interference rejection of the antenna system to be tested. The field trials focused on receive (uplink) beamforming. The system was based on the DCS-1800 (GSM) air interface and the base station arrangement incorporated a calibration subsystem for on-line calibration of the receiver and transmitter chains' amplitude and phase mismatches. The GSM *broadcast control channel* (BCCH) was transmitted on a separate frequency using a 120° sector antenna and provided a pilot signal for the mobile station as well as a reference to compare the performance of the array system against.

6.5.1.1 Field trial location and data logging. The TSUNAMI II equipment was installed on the Orange PCS test-bed in Bristol (UK). Here both Macro and Micro deployments of the smart antenna facet were investigated using mounting heights of 27 m and 8 m respectively. In the following sections, a sub-set of the results obtained from the macro-cellular trials with the array bore-sight facing due East (±1°) are described. In order to ascertain the relative performance of the candidate algorithms, the following information was logged during the field trials:
At the mobile

- The signal strength (RxLev), downlink quality metrics (RxQual) and timing advance.

- The GPS location of the wanted mobile.

At the base station

- The estimated uplink signal strength received at each element of the array, and the beamformer output.

- The call quality metrics (RxQual) for the uplink (wanted mobile to base station).

- The direction of arrival (DoA) information produced by the adaptive beamforming algorithm, and the transmit and receive weight vectors synthesised by the adaptive processing.

6.5.1.2 User tracking. The ability to track mobile users is a fundamental requirement of smart antenna technology and was appraised by means of drive tests using numerous routes in North Bristol. The performance of each candidate control algorithm to track a user's location, or *direction of arrival* (DoA) was assessed by means of deriving the angle of maximum directivity after adaptation and comparing this with position location obtained via a GPS receiver.

In order to make the user tracking test scenario emulate SFIR operation, an interference source generating a continuous wave tone at a 10 kHz offset from the traffic channel frequency was emitted from a stationary vehicle located close to bore-sight at 6 km from the cell site. The output power of the interference source for these tests was chosen by initially driving the test route and maintaining a call in the absence of any uplink interference (see figure 6.8). The uplink AGC settings for this test were observed and the interference level then adjusted to provide a *carrier-to-interference ratio* (CIR) with a dynamic range of -10 to $+27$ dB over the test route of interest. In the following results the mobile station followed the route from Rudgeway to Winterbourne via Earthcote, with the interference source located at Iron Acton. This range is by no means a worst-case scenario, but allowed calls to be maintained for sufficient time to compare the algorithms (as the interference was static and does not change significantly between the tests analysed). When operating in the presence of uplink interference it was

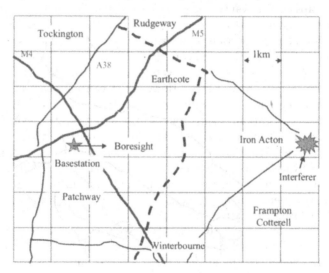

Fig. 6.8 Map of field trial site at Almondsbury, Bristol (UK). – – – – – vehicle track. Source: M. A. Beach, C. M. Simmonds, P. Howard, P. Darwood, *European Smart Antenna Test-Bed - Field Trial Results*, IEICE Trans on Comms, Vol E-84B, No 9, pp 2348–2356, Sept, 2001.

expected that some of the control algorithms would incorrectly assign the DoA of the interference signal as the DoA of the wanted mobile station (MS). This comment particularly applies to the basic algorithms (MRC and GoB), which assign the MS DoA as the direction of maximum signal strength. Thus, in negative CIR conditions, the estimated MS DoA can be expected to be that of the interference signal, giving rise to potentially large DoA errors. This can lead to a substantially reduced beamformer CIR and possible call failure (see section 6.5.1.4). Furthermore, algorithms that estimate the MS DoA with some added 'intelligence' are also subject to errors, although the DoA error was greatly reduced in these cases. For the optimal combining and AUC algorithms, a cause of DoA error is the imperfect cross-correlation of the GSM training sequence of the received user's signal and the reference signal. In the case of the MUSIC algorithm, the wanted signal is assigned as the initial signal received at the beamformer, with all others being assigned as interferers. Thus, once the tracking algorithm has become confused (for instance, wanted and interference signal not sufficiently angularly separated), incorrect signal assignment can result in a DoA error, and in the worst case the MS can be tracked (and nulled) as an interferer.

Figure 6.9 shows the estimated wanted user and GPS DoAs for the maximal ratio combining (MRC), grid of beams (GoB), optimal combining (OPT) and the Aalborg (AUC) algorithms. The approximate interference DoA is also shown (+4°). Gaps in the wanted user DoA estimate represent periods when the call was not maintained. As can be seen, for the MRC and GoB curves the MS DoA has a substantial proportion of its values on (or around in the case of GoB) the interference location, thus indicating an incorrect main beam assignment. As explained above this is to be expected in negative CIR conditions, where the mobile is in a fade for instance. However, it can also be seen that the optimal combining algorithm has a significant proportion of its MS DoA estimates around the interference location. This is the result of imperfect cross-correlation properties of the wanted and reference signal. The AUC algorithm, which also uses the GSM training sequence (to select a main beam direction), has only a small proportion of its measurements allocated to the incorrect signal DoA. This is principally due to the averaging employed in the algorithm and thus, in this particular interference scenario, represents an improvement over the optimal combining algorithm in terms of reduced DoA error.

In the case of the MUSIC algorithm, the Kalman filtering employed in the tracking allows the tracking of more than one signal source. It can be seen from figure 6.10, that until the angular separation between the wanted and interferer signals is less than approximately 8°, the algorithm successfully tracks both signals. However, once the angular separation is too small, the algorithm becomes confused, leading eventually to the MS being tracked (and thus nulled) as an interferer.

The DoA error statistics, principally resulting from incorrect MS and interference DoA assignment as discussed above, are displayed in table 6.1. In the case of no uplink interference (test 071) the DoA estimate is very good, as expected with accurate receiver chain calibration. However, for the MRC, GoB and OPT trials, upon the introduction of deliberate uplink interference, the root-mean-squared (RMS) and standard deviation (STD) of the DOA error increases substantially. This indicates large periods of the tests where the main beam is pointing to the incorrect (i.e., interference) DoA, as shown in figure 6.9. This is confirmed further with the reduction in the RMS and STD of the DoA error in the case of the AUC algorithm. For the MUSIC algorithm, only the portion of the trial before the wanted user track diverges from the true trajectory was used in the DoA error statistics, as after this point (225 s) the tracking algorithm was confused and the statistics are meaningless.

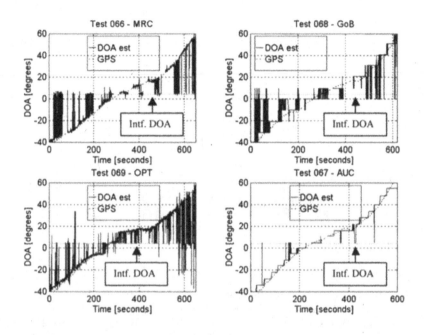

Fig. 6.9 MS and GPS DoA estimates for uplink interference trials. Source: M. A. Beach, C. M. Simmonds, P. Howard, P. Darwood, *European Smart Antenna Test-Bed - Field Trial Results*, IEICE Trans on Comms, Vol E-84B, No 9, pp 2348–2356, Sept, 2001.

6.5.1.3 Bearer quality assessment. In addition to the tracking performance of the candidate algorithms discussed above, the *quality of service* (QoS) of the traffic channel bearer was also recorded during the TSUNAMI II trials [140]. Figure 6.11 to figure 6.15 gives the recorded GSM received signal quality parameter 'RxQual' for both uplink (UL) and downlink (DL) as well as 'RxLev' of the downlink as a function of the angular track of the mobile station as observed from the base station site. Drive tests (see figure 6.8) were conducted for each control algorithm in turn with a stationary interferer located just off boresight, again approximately 6 km from the base station. For each test the interfering signal was maintained 'ON', as indicated by the grey shaded areas in the figures, until the call was dropped due to poor QoS of the bearer. The interference was then switched 'OFF' in order to re-establish the call.

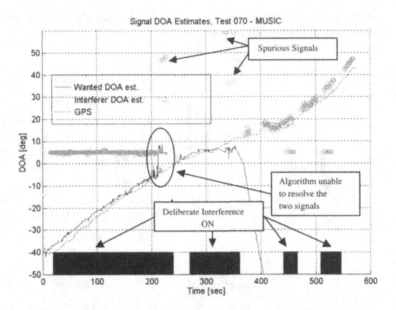

Fig. 6.10 Mobile station interference and GPS DoA estimates for MUSIC. Source: M. A. Beach, C. M. Simmonds, P. Howard, P. Darwood, *European Smart Antenna Test-Bed - Field Trial Results*, IEICE Trans on Comms, Vol E-84B, No 9, pp 2348–2356, Sept, 2001.

It can be seen from figure 6.11 that when using MRC the traffic bearer could only be maintained when 'RxLev' was high and calls were frequently dropped when the interfering signal was present. Further, the quality of the call was generally poor from both measurements and the subjective quality reported by the field trial team. Array control based on GoB (see figure 6.12) was able to maintain a bearer for a longer period when compared to the MRC results. However, there is still a large link outage and the subjective quality of the link was poor. The optimum combining approach, as given in figure 6.13, was able to maintain the link for the full duration of the trial. Here, the average link quality was below that expected for GSM (3.2), however an acceptable subjective quality audio report was recorded until the wanted mobile station and interferer were coincident within the azimuth plane of the array. Figure 6.14 shows the bearer quality recorded for the spatial reference beamforming approach using the combination of MUSIC and a Kalman filter in order to determine the parameters of the beamformer. It can be seen that initially the call quality is good, but as the wanted mobile approaches the interferer the call quality falls and tracking becomes unreliable (see figure 6.10), with a considerable link outage. Finally, figure 6.15 shows the

Table 6.1 DoA error statistics (degrees). *Note 1:* No interference present. *Note 2:* Only the portion of the test before the wanted track diverged is considered (<350 s). Source: M. A. Beach, C. M. Simmonds, P. Howard, P. Darwood, *European Smart Antenna Test-Bed - Field Trial Results*, IEICE Trans on Comms, Vol E-84B, No 9, pp 2348–2356, Sept, 2001.

	MRC 071 (Note 1)	MRC 066	GoB 068	OPT 069	MUS 070 (Note 2)	AUC 067
Mean	+1.1	+1.0	+2.0	-1.4	+1.9	+0.2
RMS	2.3	12.5	10.5	12.9	3.1	6.7
STD	2.0	12.4	10.3	12.8	2.4	6.7

performance observed using the AUC algorithm. Good call quality is maintained through the drive test, except when the signal separation is less than 10.

Table 6.2 provides a summary of the results obtained from the bearer quality drive tests in the presence of a stationary interference source. In the table, two different angular segments have been considered in terms of wanted mobile's location and that of the stationary interferer. Noting that RXQUAL of '0' relates to an excellent bearer quality, a value of '7' indicates an unusable service, the AUC algorithm provides the best QoS observed, since MUSIC only provided a link for 50% of the trial duration. Further, neither MRC and GOB algorithms were able to provide a link when the separation between source and interferer was less than 15°. Clearly, the algorithms that exploit the correlation with a known reference sequence, eg. OPT and AUC, provide enhanced performance when compared to MRC, GOB and MUSIC.

6.5.1.4 Range extension. Adaptive antennas are able to offer an increase in antenna gain when compared to a standard sectorised antenna by reducing the beamwidth of the radiation pattern from the antenna. A reduction in the angular coverage of the antenna results in an increase in the effective antenna gain in the chosen direction. Full sector coverage can then be achieved by adaptively changing the beam direction to track the mobile. Antenna gains achieved by this method are valid for both uplink and downlink and can be used to enhance the range of the base station. It is also possible to obtain a range extension in a cellular system by increasing the antenna gain using other methods. One such example is increasing the gain of a standard sectored antenna. This can only be achieved at the expense of the vertical beamwidth of the antenna and places practical limits on the design of such antennas. Commercially available antennas are not readily available with gains in excess of 19 dBi. Sector splitting is another example of achieving a range enhancement by reducing the angular coverage of each sector and benefitting from

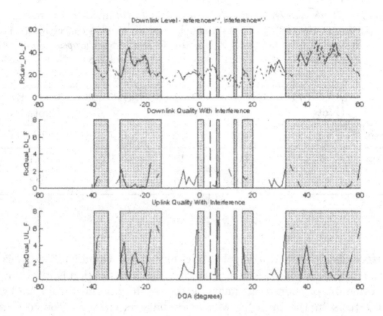

Fig. 6.11 Stationary interference QoS recording for maximal ratio combining (MRC). Source: M. A. Beach, C. M. Simmonds, P. Howard, P. Darwood, *European Smart Antenna Test-Bed - Field Trial Results*, IEICE Trans on Comms, Vol E-84B, No 9, pp 2348–2356, Sept, 2001.

the increased antenna gain as a consequence of the narrow beamwidths. Increasing the number of sectors at a base station site requires an increase in the number of transceivers and thus compounds the problem of frequency planning by requiring several closely placed sectors.

To achieve range extensions without using high gain antennas requires an improvement in the sensitivity of the base-station receiver along with an increase in base-station transmit power. It is necessary to improve both the uplink and downlink link budgets with modifications to the base-station, as it is not possible to change the mobiles that are already deployed. Hardware modifications to the base-station are relatively expensive and the sensitivity of the receiver can only be improved by a few dB. The solution to range extension offered by adaptive antennas avoids many of the pit falls that exist for the alternatives out-lined above. The TSUNAMI (II) testbed system was used to quantify the usable range extension offered by an adaptive antenna system, in a real cellular deployment. Here, the macrocellular test route shown in figure 6.8 was extended in a north-easterly direction in order to appraise the potential benefits.

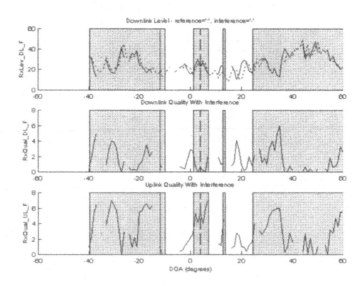

Fig. 6.12 stationary interference QoS recording for grid of beams (GoB). Source: M. A. Beach, C. M. Simmonds, P. Howard, P. Darwood, *European Smart Antenna Test-Bed - Field Trial Results*, IEICE Trans on Comms, Vol E-84B, No 9, pp 2348–2356, Sept, 2001.

The theoretical gain of an eight-element array is 9 dB relative to the single-element case. Assuming 3.5 dB/decade path loss for the macrocelluar service area, this equates to a potential range extension of 81%. To determine the effective antenna gain of the physical system, the received powers before and after the beamformer were analysed for a number of test runs, as summarised in table 6.3. The algorithms that rely on beamsteering, fall short of achieving the theoretical gain by 1 dB, to offer a maximum gain of 8 dB. This suggests that the uplink calibration of the system was not ideal, since calibration updates were only manually applied on several occasions per day during these measurements. Optimal combining comes very close to achieving the 9 dB maximum gain for the majority of the trials, however, this algorithm is less stable than MUSIC or grid of beams and exhibits regions of very low gain. Also shown in the table is the variation in single element (SGL) gain across the array. Extrapolating from the mean gain (7.4 dB) observed in these trials, the predicted range extension is 62% (again assuming a 3.5 dB/decade path loss).

Rather than assuming an idealised path-loss model, the range extension of the uplink was measured directly in the field experiments. This approach was adopted since the output power of the mobile station is limited to a maximum of 1 W, whereas the transmit power of the base station can be readily increased, thereby

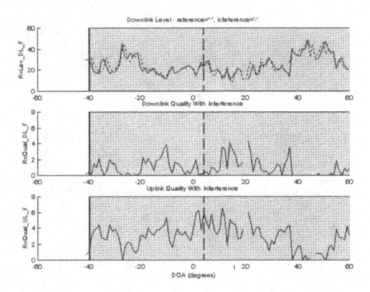

Fig. 6.13 Stationary interference QoS recording for optimum combining (OPT). Source: M. A. Beach, C. M. Simmonds, P. Howard, P. Darwood, *European Smart Antenna Test-Bed - Field Trial Results*, IEICE Trans on Comms, Vol E-84B, No 9, pp 2348–2356, Sept, 2001.

extending the downlink range. Table 6.4 gives the achievable range of the cell site for both non-adaptive (single element case) and the adaptive array modes of operation, as well as considering different bit error rate (BER) thresholds. For BER thresholds of <5% and <1.5%, maximum range extensions of 54% and 85% against the single-element case were observed respectively, with the lower BER threshold yielding the greatest gain. The results presented here only consider range extension of the traffic channel, however, the coverage of the BCCH carrier must also be increased in order to support fully GSM functionality. Given that this channel must have a fixed coverage pattern (non-adaptive), then the transmit power of this bearer must be increased in order to fully support the enhanced range of the traffic channels.

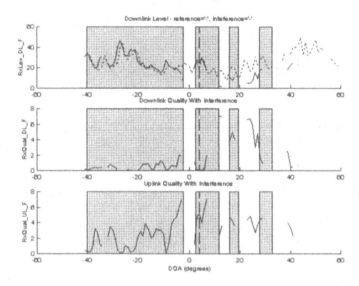

Fig. 6.14 Stationary interference QoS recording for MUSIC (MUS). Source: M. A. Beach, C. M. Simmonds, P. Howard, P. Darwood, *European Smart Antenna Test-Bed - Field Trial Results*, IEICE Trans on Comms, Vol E-84B, No 9, pp 2348–2356, Sept, 2001.

6.5.2 Downlink Beamforming for Third-Generation Mobile Wireless Networks

The following discussion is based upon reference [7] which presents a comprehensive analysis of the application of switched beam beamformers to the downlink of third-generation mobile wireless systems (3G).

3G mobile communications systems have been developed to supersede GSM (2G) systems and provide enhanced services to the others. Two such services are:

- **Multimedia streaming.** This provides video and audio to the user's handset on demand; and

- **Location-based services.** This allows the user to be provided with information based upon the locality of the user, e.g., finding a restaurant.

The standard for such systems has been referred to as the *Universal Mobile Telephone System* or UMTS, and the wireless section of the standard is referred to as *UMTS Terrestrial Radio Access* or UTRA. UTRA is based upon direct sequence CDMA technology and the basic configuration for the FDD based standard is given in table 6.5 [141].

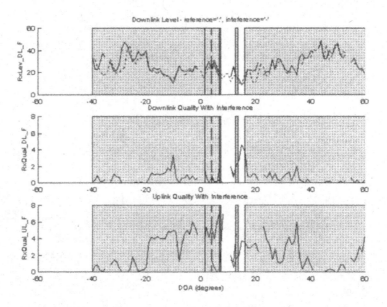

Fig. 6.15 Stationary interference QoS recording for temporal reference grid of beams (AUC). Source: M. A. Beach, C. M. Simmonds, P. Howard, P. Darwood, *European Smart Antenna Test-Bed - Field Trial Results*, IEICE Trans on Comms, Vol E-84B, No 9, pp 2348–2356, Sept, 2001.

'3G' mobile wireless technology is based on direct sequence code division multiple access (DS-CDMA). The signal bandwidth is 5 MHz, which is larger that the coherence bandwidth of the transmission channels. Consequently, the signal is subject to wideband propagation, i.e., frequency selective fading is observed at the receiver. Thus the term *wideband CDMA* is used for the technology. Data rates of up to 384 kb/s have been predicted and location awareness of the users is a legal requirement for operation of any mobile telephone system in Europe and North America. The standard for '3G' is referred to as *Universal Mobile Telecommunications System* or UMTS. This covers the complete system including the wireless section and core network. The wireless section of the standard is called *UMTS Terrestrial Wireless Access* or *UTRA*. Duplex transmissions are made using frequency division duplex (FDD) and the uplink and downlink channels are separated by around 120MHz.

Transmit, or downlink beamforming, provides additional challenges to the system designer when operating an FDD system which have been described in chapter 5. Switched beam antennas, which have been described in chapter 4, are an alternative to the fully adaptive, computational intensive downlink beamforming

Table 6.2 'RxQual' as a function of angular separation (— indicates that a link was not available). Source: M. A. Beach, C. M. Simmonds, P. Howard, P. Darwood, *European Smart Antenna Test-Bed - Field Trial Results*, IEICE Trans on Comms, Vol E-84B, No 9, pp 2348–2356, Sept, 2001.

Algorithm	Uplink qual. -20° to -15°	Uplink qual. -15° to -10°	Downlink qual. -20° to -15°	Downlink qual. -15° to -10°
MRC	3.4	—	1.0	—
GoB	4.7	—	1.6	—
OPT	3.2	4.1	1.1	2.4
MUS	2.5	2.9	0.5	0.9
AUC	3.2	4.2	0.7	1.3

Table 6.3 Antenna gains measured in the range extension trials. Source: M. A. Beach, C. M. Simmonds, P. Howard, P. Darwood, *European Smart Antenna Test-Bed - Field Trial Results*, IEICE Trans on Comms, Vol E-84B, No 9, pp 2348–2356, Sept, 2001.

Mode	Mean gain	Peak gain
SGL	0.2 dB	1.0 dB
GoB	7.1 dB	8.0 dB
OPT	8.2 dB	9.0 dB
MUS	7.4 dB	8.0 dB
AUC	7.2 dB	8.0 dB

approaches that are necessary when operating over large FDD frequency offsets. The system architecture is identical to that of an uplink switched beam system such as that shown in figure 6.16. Tirrola and Ylitalo have appraised a system using two beams over a 120° sector based on a two-port Butler matrix [131]. In the reference, the downlink beam is selected corresponding to the uplink beam that gives the highest received signal strength. Results show a performance improvement for channels where the angular spread is considerably less than the beamwidth. The authors also conclude that two beams per sector is not sufficient as cusping loss impairs service to users located in this region.

Table 6.4 Range extension BER versus algorithm. Source: M. A. Beach, C. M. Simmonds, P. Howard, P. Darwood, *European Smart Antenna Test-Bed - Field Trial Results*, IEICE Trans on Comms, Vol E-84B, No 9, pp 2348–2356, Sept, 2001.

Mode	Distance BER < 1.5%	BER < 5%
SGL	7 km	11 km
GoB	13 km	15 km
OPT	13 km	15 km
MUS	8 km	13 km
AUC	13 km	17 km

Table 6.5 UTRA FDD System Parameters.

Parameter	Value
European frequency allocation	Uplink: 1920 MHz − 1980 MHz) Downlink: 2110 MHz − 2170 MHz)
FDD Offset	Minimum: 134.5 MHz Maximum: 245.2 MHz Nominal requirement: 190 MHz
Chipping rate	3.84 Mcps
Spreading factor	4 − 512
Nominal RF bandwidth	5 MHz ± 200 kHz
Modulation	QPSK
Scrambling code	Uplink: Kasami (short), Gold (long) Downlink: Gold

The key advantages of these switched systems over fully adaptive solutions are:

- *Inherent numerical stability.* It does not require intense signal processing since beam selection can be based on RF, analogue signal processing and simple control logic. Consequently, algorithm robustness is not usually a design issue.

- *Considerable reduction in computational overhead and delay.* As stated above, complex signal processing algorithms are not required.

- *Simple implementation.* These systems can be retrofitted to existing cell sites without the complexity associated with digital or adaptive beamforming schemes which require calibration.

Furthermore, a minimum isolation between co-channel users of around 13 dB is achievable (indicated by the sidelobe levels of the beam pattern), whereas fully adaptive techniques can suffer from poor null depth or high sidelobes due to angular spread and non-ideal array calibration. Hence, there is no 'degree of freedom' limitation in switched beam systems when subject to high user densities. In what follows, a simple closed-form expression for evaluating the capacity

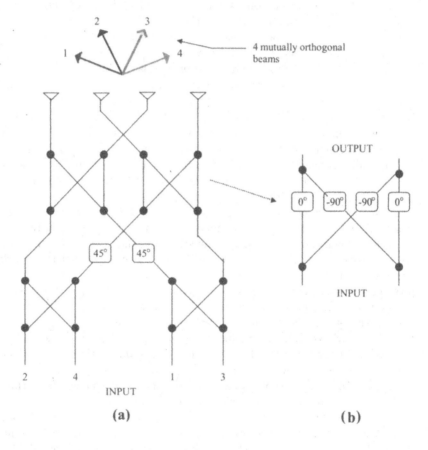

(a)   **(b)**

Fig. 6.16 Butler matrix. (a) 4 × 4 Butler matrix. (b) A hybrid used to form a Butler matrix. Source: B. Allen, M. Beach, *On The Analysis of Switch-Beam Antennas for the W-CDMA Downlink*, IEEE Trans. Veh. Tech., Vol 53, No 3, pp 569–578, May, 2004 © IEEE.

increase of W-CDMA networks employing switched-beam antenna systems is described. The expression incorporates the effect of practical antenna patterns and the impact of multipath scattering on code orthogonality, as well as that of pilot power signal. The relevant terminology is now defined.

Loss of code orthogonality (α). This is either caused by time dispersion (as a result of multipath propagation), non-orthogonal code sets, or asynchronous signals arriving at the receiver [142, 143], as well as a combination of these effects. This directly impacts the level of multiple access interference in terms of self-interference. Here, $\alpha = 0$ represents complete code orthogonality and $\alpha = 1$ represents a complete loss of orthogonality.

Beam efficiency (ξ) accounts for power spilled into adjacent beams which increases co-channel (inter-beam) interference. This is caused by the magnitude of cusping loss between beams, sidelobe levels and angular spread in the channel and takes a value of between 0 and 1, where 1 represents complete efficiency. The impact of each of the terms is shown later in this section.

Pilot signal power is the proportion of power allocated to the pilot signal ($1 - \beta$, where β is power allocated to the users). The pilot signal is used to transmit the Broadcast Control CHannel (BCCH) and Paging CHannel (PCH) [144]. It is allocated a separate code and is common to all users and thus is transmitted on a fixed coverage antenna within the entire cell or sector. The pilot channel signal orthogonality is therefore equivalent to that of an omni-sector transmission.

Traffic loading fraction (γ) represents the net downlink traffic load where $\gamma = 1$ is the fully loaded condition.

Intra-cell path-loss (Lp_{Intra}) models the total path-loss experienced by transmissions from the user's own base station as a function of distance.

Inter-cell path-loss, (Lp_{Inter}), is the sum of path-losses from surrounding base stations to the user. Here, a cell radius of 1 km, a three-tier network with a single cell reuse distance and uniform user distribution has been used in these investigations, i.e., 18 interfering base stations, together with a power-law path-loss model employing a path-loss exponent of $n = 4$, which is deemed to be appropriate with a base station height of 10 m and mobile height of 2 m. These parameters are in line with W-CDMA system appraisals reported in [142] and [143].

Shadowing (χ) is modelled as a log-normal random variable with variance σ_{SH} that models the shadowing experienced by the user [101].

A closed-form expression can now be developed that enables the performance of switched beam antennas to be ascertained. The equation incorporates the terms defined above. The form of the equation considers the total transmit power from each base station and, since the signals pass through a common channel on the downlink, power control to each individual user can be considered as ideal in this analysis.

The downlink inter-cell interference received by a user in a network of J interfering basestations with omni-sector directional antennas is given by equation (6.5). This assumes long-term temporal averaging of the received signal since fast fading effects have been omitted.

$$I_{InterOM} = \chi_{InterOM} \sum_{j=1}^{J} L_{pInterj} \cdot \tag{6.4}$$

$$(\beta_j \cdot \gamma_j \cdot \alpha_{InterOMj} + (1 - \beta_j) \cdot \alpha_{InterOMj}) \tag{6.5}$$

where $\alpha_{InterOMj}$ represents the inter-cell orthogonality factor with omni-sector antennas at the jth interfering base station and $\chi_{Inter_{OM}}$ represents the instantaneous shadowing coefficient resulting from the J interfering base stations when operating with omni-sector antennas.

Similarly, intra-cell interference can be computed with omni-sector antennas using equation (6.7). Also, the inter- and intra-cell interference for cells using switched beam antennas are computed from equations (6.9) and (6.11) respectively. Note that the $(1 - \beta)$ term in the equations represents the pilot power fraction that does not reduce as a function of beamwidth since it is required to be transmitted simultaneously across the entire coverage area.

$$I_{IntraOM} = \chi_{IntraOM} L_{pIntra} \cdot \tag{6.6}$$

$$(\beta \cdot \gamma \cdot \alpha_{IntraOM} + (1 - \beta) \cdot \alpha_{IntraOM}) \tag{6.7}$$

$$I_{InterSB} = \chi_{InterSB} \sum_{j=1}^{J} L_{pInterj} \cdot \tag{6.8}$$

$$\left(\frac{\beta_j \cdot \gamma_j \cdot \alpha_{InterSBj}}{N_j \cdot \xi_j} + (1 - \beta_j) \cdot \alpha_{InterSBj} \right) \tag{6.9}$$

$$I_{IntraSB} = \chi_{IntraSB} L_{pIntra} \cdot \tag{6.10}$$

$$\left(\frac{\beta \cdot \gamma \cdot \alpha_{IntraSB}}{N_j \cdot \xi_j} + (1 - \beta) \cdot \alpha_{IntraSB} \right) \tag{6.11}$$

Where $\alpha_{IntraSB}$ = intra-cell orthogonality factor with switched beam antennas. $\alpha_{IntraOM}$ = intra-cell orthogonality factor with omni-sector antennas. $\alpha_{InterSB}$ = inter-cell orthogonality factor with switched beam antennas. $\alpha_{IntraPilot}$ = intra-cell orthogonality factor of the omni-sector pilot channel. $\alpha_{InterPilot}$ = inter-cell orthogonality factor of the omni-sector pilot channel. $\alpha_{IntraSB}$ = shadowing coefficient from the user's base station operating with a switched beam antenna. $\alpha_{IntraOM}$ = shadowing coefficient from the user's base station operating with an omni-sector antenna. $\alpha_{InterSB}$ = shadowing coefficient from the interfering base stations operating with switched beam antennas. $\alpha_{InterOM}$ = shadowing coefficient from the interfering base stations operating with omni-sector antennas N = number of equal beamwidth beams.

Thus, the *interference reduction ratio* (IRR) obtained by employing a switched beam antenna compared to the single omni-sector case is given by equation (6.12). In interference limited networks, a reduction in interference can be translated into a capacity gain. Hence, equation (6.12) also gives an indication of the capacity gain obtainable from switched beam antennas. The equation compares the total interference power received by the user in a network employing omni-sector antennas with that of a network employing switched beam antenna systems in each sector and assumes the received powers combine linearly.

$$IRR = \frac{I_{InterOM} + I_{IntraOM}}{I_{InterSB} + I_{IntraSB}} \qquad (6.12)$$

Now, numerical results are obtained for the expression developed above. The downlink IRR observed by employing a switched beam antenna in the central cell compared to a 120° fixed coverage omni-sector case. The expression can be evaluated as a function of the following system parameters:

- Number of beams; and

- Beam efficiency.

The expressions also allow the following parameters to be easily evaluated, and results relating to these are found in [7]:

- traffic loading;

- pilot power fraction;

- path loss;

- code orthogonality;

- shadowing variance; and

- shadowing correlation.

Here, performance as a function of the number of beams and beam efficiency are presented. Results relating to the other parameters can be found in [7].

6.5.2.1 IRR versus number of beams and beam efficiency. This evaluation is initially carried out using the system parameters given in table 6.6, where the number of beams is incremented from 1 (omni-sector) to the extreme case of $N = 32$. The path-loss values have been chosen to represent a user located midway between the base station and cell boundary using the path-loss model stated earlier. The shadowing coefficients, χ, are set to 0 dB for this investigation. A representative value of pilot power fraction is specified in [144], however, actual values are currently unavailable in the literature. For this investigation, the

Table 6.6 W-CDMA system test parameters. Source: B. Allen, M. Beach, *On The Analysis of Switch-Beam Antennas for the W-CDMA Downlink*, IEEE Trans. Veh. Tech., Vol 53, No 3, pp 569–578, May, 2004 © IEEE.

Parameter	Value
Number of beams (N)	1−32
Traffic loading factor (γ)	100%
Pilot signal power fraction ($1 - \beta$)	20%
$\alpha_{InterOM}$	1
$\alpha_{InterSB}$	1
$\alpha_{IntraOM}$	0.4
$\alpha_{IntraSB}$	See table 6.7
Beam efficiency (ξ)	See table 6.7
Total inter-cell path-loss(Lp_{Inter})	90 dB
Intra-cell path-loss (Lp_{Intra})	80 dB
$\chi_{InterSB}$, $\chi_{IntraOM}$, $\chi_{IntraSB}$, $\chi_{InterOM}$	0 dB

network is assumed to be fully loaded ($\gamma = 1$). $\alpha_{InterOM} = 1$ is assumed since W-CDMA utilises an asynchronous air interface, hence inter-cell codes will not be synchronised rendering the possibility of worst-case cross-correlation properties between them. Since this value is due to asynchronous codes and not time dispersion, employing a narrower beam will not reduce $\alpha_{InterSB}$. The value of $\alpha_{IntraOM}$ for an omni-sector antenna in table 6.6 has been computed in [144] and is representative of a macro-cell with a 120° antenna and the derivation of the remaining values is now described.

The primary mechanism for orthogonality loss is, in this case, time dispersion [142]. By reducing the beamwidth, an improvement in orthogonality may be expected and is estimated in table 6.7 for various beamwidths (Bw) assuming a chipping rate of 3.84 Mcps, basestation antenna mounted above rooftop height and the scatterers modelled as a ring of radius $L/2$ around the user (where L is the system's range resolution given by $L = c \cdot T_c$, and the speed of light $c = 3 \times 10^8$ m/s) shown in figure 6.17. The resulting value is estimated by equation (6.13).

$$\alpha_{IntraSB} = \frac{\alpha_{IntraOM} \cdot Bw}{120°} \qquad (6.13)$$

The loss of orthogonality is manifested when the propagation delay between two impulses arriving at the receiver is less than the chip period, T_c. Consequently, the rake receiver cannot resolve the paths and the fading that results, which causes interference at the detector output. The delay corresponds to a path-length difference of $\Delta d \leq L$. In the case of W-CDMA, $T_c = 260.4$ ns and therefore

Table 6.7 W-CDMA switched beam test parameters. Source: B. Allen, M. Beach, *On The Analysis of Switch-Beam Antennas for the W-CDMA Downlink*, IEEE Trans. Veh. Tech., Vol 53, No 3, pp 569–578, May, 2004 © IEEE.

N	Ideal Azimuth 3 dB Beamwidth (Deg)	$\alpha_{IntraSB}$	Beam efficiency (%)
Omni-sector	120	0.4	90
4	30	0.2	90
8	15	0.1	90
16	7.5	0.05	65
24	5	0.025	40
32	3.75	0.0125	27.5

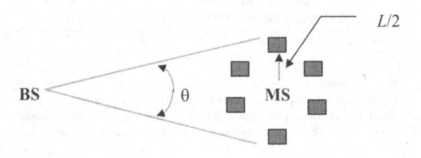

Fig. 6.17 Ring of scatterers. Source: B. Allen, M. Beach, *On The Analysis of Switch-Beam Antennas for the W-CDMA Downlink*, IEEE Trans. Veh. Tech., Vol 53, No 3, pp 569–578, May, 2004 © IEEE.

$L = 78$ m. Scatterers within the area defined by d, d_1 and d_2 in figure 6.18 cause these path-length differences and scatterers outside the area will cause path-length differences of $\Delta d > L$ which are resolvable by the rake receiver. For a mobile station located at $d = 500$ m from the basestation, the model in figure 6.18 yields $d_1 = 145$ m and $d_2 = 39$ m. By reducing the antenna beamwidth, the number of scatterers illuminated inside the locus will reduce but this can only occur when the beamwidth significantly reduces coverage within the locus. Assuming uniformly distributed scatterers throughout the coverage area, the required beamwidth (θ) to reduce the number of scatterers by 50% is given by equation (6.14) and yields $\theta = 30°$ for the above example. As a consequence of reducing the scattering within the locus, the number of un-resolvable multipaths is reduced, thus helping to preserve orthogonality. Note, however, that by employing narrow beamwidth

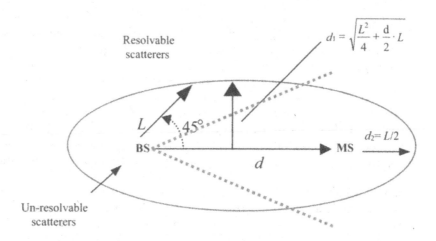

Fig. 6.18 Locus of scatterers causing orthogonality loss. (..............30°*beamwidth*).
Source: B. Allen, M. Beach, *On The Analysis of Switch-Beam Antennas for the W-CDMA Downlink*, IEEE Trans. Veh. Tech., Vol 53, No 3, pp 569–578, May, 2004 © IEEE.

antennas the resolvable multipaths present at the rake receiver are also reduced which will cause the achievable path diversity gain to diminish. A beamwidth of $\theta = 30°$ is achievable with a uniformly weighted four-element linear array.

$$\theta = 2 \cdot \tan^{-1} \left[\frac{2 \cdot \sqrt{\frac{L^2}{4} + \frac{d_1}{2}} L}{d_1} \right] \tag{6.14}$$

Using the above analysis, the orthogonality factors are given in table 6.7 for a number of beamwidths. The table also shows the beam efficiencies that have been used in this investigation. Three parameters contribute to beam efficiency: radiation pattern loss (L_{Rad}), beam cusp loss (L_{Cusp}) and angular spread loss (L_{AS}). The cumulative effect is given by equation (6.15).

$$\xi = (1 - L_{Rad}) + (1 - L_{Cusp}) + (1 - L_{AS}) \tag{6.15}$$

Radiation pattern loss (L_{Rad}) is defined by equation (6.16) [9] and describes the power radiated through the main beam compared to the total radiated power. The theoretical 'top hat' antenna pattern that does not possess sidelobes would therefore yield $L_{Rad} = 1$.

$$L_{Rad} = 1 - \frac{P_{Main}}{P_{Total}} \tag{6.16}$$

P_{Main} is the power transmitted in the main beam (defined as the area between the first two nulls) and P_{Total} is the total transmitted power. The above definition considers a three-dimensional radiation pattern but is equally valid for the azimuth pattern assuming a uniform radiation pattern in elevation. The azimuth radiation pattern for an eight-element (ULA) with $\lambda/2$ element spacing shown in figure 6.19 yields L_{Rad} to be 90% which has been used throughout this analysis.

The presence of adjacent beams overlapping with the wanted main beam will further degrade the beam efficiency (L_{Cusp}). A cusping loss of 4 dB is a typical figure (although it can be reduced as reported in [31]). The impact of cusping loss on system performance depends upon the angular position of the mobile station. In order to optimise beam efficiency, it is desirable to minimise L_{Cusp}. This would, however, compromise quality of service to users located in the cusp region and therefore highlights a design trade-off. For simplicity, cusping loss is assumed ideal ($L_{Cusp} = 0$) in this analysis, thus further analysis is required to fully model the impact of cusping loss. Once the beamwidth becomes smaller than the largest angular spread, this latter effect dominates, causing energy to spread out of the main beam with the possibility of increasing interference levels to co-channel users. This *residue beamwidth* (RB) is illustrated in figure 6.20 and can be defined as the *angular spread* (AS) observed when an infinitely narrow beam excites the channel. Figure 6.20 shows that the beamwidth is limited by the angular spread. This is reflected in table 6.7 where beam efficiency rapidly reduces for beamwidths of less than the angular spread. Throughout this investigation an angular spread of 10° is considered representative of an urban environment and falls within the range reported in [99]. L_{AS} is computed by equation (6.17) where BW is the azimuth beamwidth and AS the angular spread.

$$L_{AS} = 1 - \frac{BW}{AS} \qquad (6.17)$$

Figure 6.21 illustrates the achievable IRR computed using the above analysis as the number of equally spaced beams within the sector is increased. Initially it is assumed that beam efficiency and $\alpha_{IntraSB}$ stay constant (at levels of 90% and 0.4 respectively) as the number of beams increases and yields the idealistic curve shown. In a practical system, the beamwidth efficiency and $\alpha_{IntraSB}$ will vary with beamwidth as shown in table 6.7. The realistic curve shown in the figure reflects a practical system performance and shows that an optimum IRR is achieved when 16 beams are used each having a 3 dB beamwidth of 7.5°. This is realised with a 16-element ULA. The figure also shows a similar performance is attainable with 8 beams of 15°. This is a more practical solution as an eight-element ULA is required therefore reducing the system cost, complexity, wind-loading and cabling associated with a larger array. Note that the figure shows little performance gain is achieved by increasing the number of beams above 16. Performance with relatively small beamwidths is highly dependent upon the level of angular spread. Here it

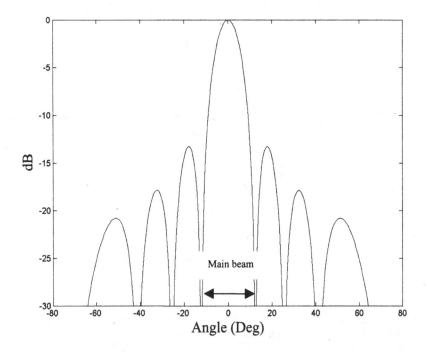

Fig. 6.19 Beam efficiency (eight-element ULA). Source: B. Allen, M. Beach, *On The Analysis of Switch-Beam Antennas for the W-CDMA Downlink*, IEEE Trans. Veh. Tech., Vol 53, No 3, pp 569–578, May, 2004 © IEEE.

has been stated that $10°$ of angular spread is assumed. Environments exhibiting higher values of angular spread will exhibit a reduction in IRR since L_{AS} will be increased. In severe cases, performance gain may be negligible.

6.5.3 User Location and Tracking

The following example is based upon reference [145]. Like the previous example, this also considers user location and tracking estimation in the context of a 3G mobile communications network.

The ability to support position location within wireless networks provides network operators with valuable services, as well as users with a host of new applications. This includes:

- navigation;

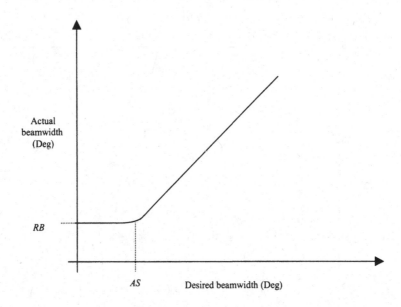

Fig. 6.20 Beamwidth Limitation Due to Residue Angular Spread. Source: B. Allen, M. Beach, *On The Analysis of Switch-Beam Antennas for the W-CDMA Downlink*, IEEE Trans. Veh. Tech., Vol 53, No 3, pp 569–578, May, 2004 © IEEE.

- location-based services;

- network management; and

- security applications.

Historically, the need for pinpointing a mobile user arises as a matter of security, in order to track emergency calls made by cellular phones (US Enhanced 911, E-911). An increasingly large fraction of E-911 calls are placed by cellular phones, which is a direct result of the growing number of cellular subscribers [146]. Location information for wireless E-911 calls permits a co-ordinated response in situations where callers are disoriented, disabled, unable to speak or do not know their location. Indeed, certain regulations adopted by the US Federal Communications Commission (FCC) require that, as from October 2001, all the emergency calls made by cellular phones must be localised to within an accuracy of 125 m in 67% of cases [147].

Although the main incentive to integrate mobile location algorithms into wireless systems is to aid emergency service call-outs, there are many other parallel applications that make positioning techniques attractive for wireless service providers. First of all, position location capabilities open the doors to

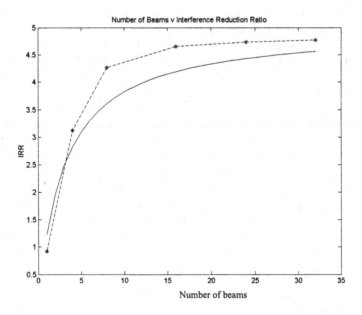

Fig. 6.21 IRR v number of beams (user located midway between BS and cell boundary. (See tables 6.6 and 6.7 for system parameters). Source: B. Allen, M. Beach, *On The Analysis of Switch-Beam Antennas for the W-CDMA Downlink*, IEEE Trans. Veh. Tech., Vol 53, No 3, pp 569–578, May, 2004 © IEEE.

a new world of previously unimagined information services that can be offered to the customer along with standard voice services. Examples of such services are: electronic yellow pages, 'where I am' applications, navigation and personalised traffic, restaurant/hotel finders. Second, position location systems enhance services in the public interest, such as fleet management, roadside assistance, traffic routing and scheduling of vehicles in real time. Real-time position location can also be used to track lost children, service personnel (police etc.), suspected criminals and stolen vehicles. This will significantly increase the crime-fighting capability of law enforcement agencies. Third, the knowledge of subscriber location is very useful for wireless service providers, because many *radio resource management* (RRM) functions utilise this additional information, favouring network design and optimisation.

A radio location system can operate by measuring and processing physical quantities related to radio signals travelling between a *mobile terminal* (MT) and a set of fixed transceivers, e.g. satellites or *base stations* (BSs).

A broad spectrum of solutions for both *modified* and *unmodified* handset solutions can be found in the open literature [146]. There are several fundamental approaches for implementing a radio location system, including those based on signal strength [148], *angle of arrival* (AoA) [149], and *time of arrival* (ToA) [150]. Each method has an associated deployment complexity, either at the terminal or network side, and an associated performance level.

Radio-location techniques are broadly classified into two categories: modified and unmodified handset solutions. The former techniques require adjustment into existing handsets, while the latter ones need modification only at the base stations or the switching centre. Modified handset solutions yield better performance than solutions that work with existing handsets [151]. The biggest drawback of these solutions, however, is the additional cost involved in re-issuing new handsets. Moreover, additional hardware will increase the weight and sizes of the mobile terminals and reduce power efficiency, which is contrary to customer desires. Installing a global positioning system (GPS) receiver on each handset, for example, seems to be the most straightforward location positioning solution. However, this option is not attractive for the following reasons:

- first of all, the GPS receiver increases weight, size, and thus the cost of the mobile terminal;

- further, GPS operates in the L-band requiring the mobile antenna to be redesigned; and finally

- the GPS receiver needs to have at least four satellites visible at all times, and this is not always guaranteed in shadowed or indoor environments.

Unmodified handset solutions require changes that involve only the system infrastructure, such as base stations or the switching centre. Although unmodified handset solutions have an inferior performance compared to modified handset solutions, they are much less expensive and do not compromise handset performance. Due to economic factors, the optimum solution to mobile location will have to rely on unmodified handsets. Here, 'optimum solution' has to be understood as a trade-off between performance gains and overall implementation complexity.

In order to reduce the implementation complexity and the signalling overhead, significant research effort has focused on developing solutions using a single BS and unmodified MTs. For instance, the solution proposed in [152] utilises information about the scatterers surrounding the BS and the position of the MT is estimated by using the Doppler shift related to each impinging path, hence requiring a moving MT. Moreover, the *fingerprint* method exploits the multipath characteristics of the signal received at the BS as a function of the MT position. According to this approach, the location of the MT is identified by matching the actual *signature* of the received signal with the entries stored in a database available at the BS. The drawback of this method is its demanding implementation, as for each BS cell extensive and accurate measurements are required for creating and updating the database. However, the fingerprint technique is gaining increasing attention for indoor applications, where database management is easier than in wide urban areas.

Here, a method for allowing a single BS antenna array to locate unmodified MT handsets is discussed. A *wideband code-division multiple-access* (W-CDMA) wireless system is considered, and it is assumed that transmission takes place over a multipath channel. The proposed algorithm uses the channel parameter estimation performed within W-CDMA receivers, namely the characteristics of the *multipath components* (MPCs) impinging on the BS in terms of their propagation delay or ToA and AoA. This naturally requires a wideband signalling scheme in combination with an array antenna at the BS and the associated parameter estimation algorithms. In addition, the method needs some minimal information about the environment in the BS neighbourhood. The location determination of a MT is then performed in two steps. First, it is decided if the MT is in *line-of-sight* (LOS) of the BS. If it is, the location is determined directly through trigonometric relationships. Second, if the MT is found to be in *non-LOS* (NLOS), then its location is determined by minimising a given *cost function*. This results in a location technique of reasonable complexity and high precision. The performance is determined for a UMTS microcellular operating scenario and a 3D deterministic channel model. The technique is summarised in the following section and full details can be found in [145].

6.5.3.1 The single base station location technique. The location technique described here is designed for microcellular wireless networks. A short introduction of dominant propagation mechanisms in microcellular environments is outlined in the following subsection which clarifies the operation of the subsequent location technique. The deployment of microcells is motivated by the desire to reduce cell sizes in areas where a large number of users require access to the cellular network. This could in theory be achieved using macrocells, where the BS antennas are placed above the average roof top height, but this would significantly increase the costs and planning difficulties. In a microcell the BS antenna is typically at the same height as lamp posts (3-6 m above ground level), although the antenna is more usually mounted on a side or corner of a building. The dominant propagation mechanisms in a microcellular environment are free space propagation, plus multiple reflection and scattering within the cell coverage area, together with diffraction around the vertical edges of buildings and over the rooftops. Since the BS antenna is always below the average roof top height, it is often surrounded by local scatterers and thus submerged into the clutter. As a result, several MPCs impinging on the BS can generally be observed, which are characterised in terms of AoA, ToA and power. Contrary to macrocells, in microcells the MPCs arrive at the BS with a large angular spread. This feature will be exploited by the proposed location technique where a micro-cellular scenario is considered in the subsequent analysis.

Prerequisites

The proposed location technique makes use of a single BS array antenna to locate unmodified MTs. Single BS solutions offer many advantages. First, the MT does not have to be synchronised with other BSs. Second, the coverage by several BSs ('*hearability*') is no longer a problem. Finally, the inter-network signalling requirement ('*back-haul*) is reduced.

The positioning technique described here utilises triangulation supported by minimal information about the clutter in the BS neighbourhood. For a given number N ($N \geq 3$) of MPCs impinging at the BS, the algorithm needs the knowledge of:

- AoAs $\alpha = [\ \alpha_1 \quad \alpha_2 \quad \dots \quad \alpha_N\ ]$; and

- absolute ToAs $\tau = [\ \tau_1 \quad \tau_2 \quad \dots \quad \tau_N\ ]$

This requires a wideband signalling scheme in combination with an array antenna at the BS and the associated channel parameter estimation algorithms.

A further prerequisite of the proposed positioning algorithm is the introduction of a *sentinel function* (SF), $\varphi(\alpha)$, $\varphi(\alpha)$, which is defined as the Euclidean distance between the BS and the nearest obstacle found along the azimuth direction identified by the angle α (figure 6.22). Only fixed obstacles lying within the cell coverage area and intersecting the azimuth plane of the BS antenna are taken into account. For implementation purposes, such a function might be sampled in steps of $\Delta\alpha$, then tabulated and eventually stored at the BS. A small sampling step, e.g. $\Delta\alpha = 1°$ or $\Delta\alpha = 0.5°$, is required when the clutter in the BS neighbourhood is quite complex. Even in this case, however, a small amount of memory is needed to store the $(360°/\Delta\alpha)$ samples at the BS. Furthermore, linear interpolation may be used to calculate $\varphi(\alpha)$ if the argument α lies between two subsequent samples. The sentinel function is the only information about the environment surrounding the BS required by the proposed location technique. A sample of a polar representation of the *sentinel function* for a BS in a typical urban environment is shown in figure 6.23. The scenario depicted in figures 6.22 and 6.23 is a district of the town of Viareggio, Italy.

LOS-vs.-NLOS decision criterion and location estimation in LOS conditions

The absolute distance travelled by the first MPC received at the BS can be calculated as

$$d_1 = \sqrt{(c\tau_1)^2 - (h_{BS} - h_{MT})^2} \qquad (6.18)$$

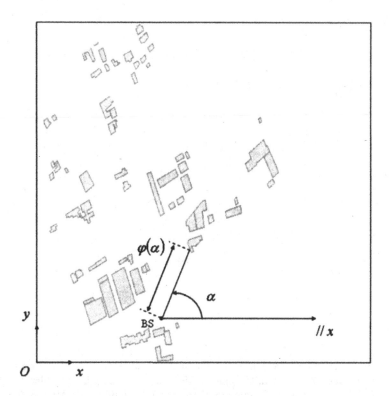

Fig. 6.22 Definition of the sentinel function $\varphi(\alpha)$ for a real urban scenario in the town of Viarrggio, Italy. Source: M. Porretta, P. Nepa, G. Manara, F. Giannetti, M. Dohler, B. Allen, A. H. Aghvami, *A Novel Single Base Station Location Technique for Microcellular Wireless Networks: Description and Validation by a Deterministic Propagation Model*, IEEE Trans. Veh. Tech., Vol 53, No 5, pp 1553–1560, Sept, 2004 © IEEE.

where τ_1 is the absolute propagation delay of the first MPC reaching the BS, h_{BS} and h_{MT} are the heights of the BS and MT, respectively. The distance d_1, the SF, $\varphi(\alpha)$, and the AoA of the first impinging MPC, α_1, are used to decide whether the MT is in LOS condition or not. If $\varphi(\alpha_1) \geq d_1$ then LOS is assumed (figure 6.24) and the position $(x\hat{}_{MT}, y\hat{}_{MT})$ of the MT is simply estimated as

$$\begin{cases} x\hat{}_{MT} = x_{BS} + d_1 \cos(\alpha_1) \\ y\hat{}_{MT} = y_{BS} + d_1 \sin(\alpha_1) \end{cases} \tag{6.19}$$

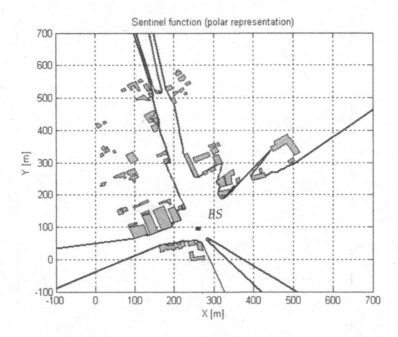

Fig. 6.23 Sample of a polar representation of the sentinel function $\varphi(\alpha)$ for a typical urban scenario. Source: M. Porretta, P. Nepa, G. Manara, F. Giannetti, M. Dohler, B. Allen, A. H. Aghvami, *A Novel Single Base Station Location Technique for Microcellular Wireless Networks: Description and Validation by a Deterministic Propagation Model*, IEEE Trans. Veh. Tech., Vol 53, No 5, pp 1553–1560, Sept, 2004 © IEEE.

Therefore, when the MT is supposed to be in LOS, only the parameters relevant to the first MPC (α_1, τ_1) are used in the location estimation. It is worth mentioning that the solution given by equation (6.19) is the basic 1−ToA/ 1−AoA single BS hybrid positioning algorithm [151]. However, such a scheme is used by the proposed location technique only if the MT is supposed to be in LOS with the BS.

Cost function minimisation in NLOS conditions

If the above mentioned test, $\varphi(\alpha) \geq d_1$, fails, then the NLOS condition is assumed (figure 6.25). In this case, the MT position is determined by minimising a given cost function. First, the algorithm evaluates the coordinates of the scatterers found along the angle of arrival for the first N MPCs, namely $\mathbf{x}_s = [\ x_{s1}, \quad x_{s2}, \quad \ldots \quad x_{sN}\ ]$ and $\mathbf{y}_s = [\ y_{s1}, \quad y_{s2}, \quad \ldots \quad y_{sN}\ ]$. These scatterers are

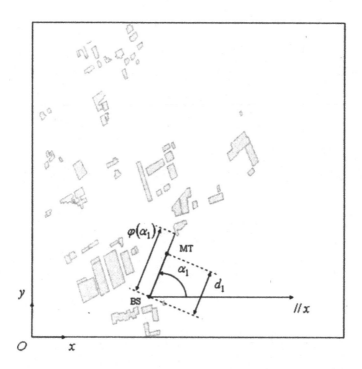

Fig. 6.24 If the mobile terminal (MT) and the BS are in LOS, then $\varphi(\alpha_1) > d_1$. Source: M. Porretta, P. Nepa, G. Manara, F. Giannetti, M. Dohler, B. Allen, A. H. Aghvami, *A Novel Single Base Station Location Technique for Microcellular Wireless Networks: Description and Validation by a Deterministic Propagation Model*, IEEE Trans. Veh. Tech., Vol 53, No 5, pp 1553–1560, Sept, 2004 © IEEE.

the points where the ray related to each MPC has been reflected or diffracted for the last time before reaching the BS; consequently, their coordinates are obtained as:

$$\begin{cases} lx_{si} = x_{BS} + \varphi(\alpha)\cos(\alpha_1) \\ y_{MT} = y_{BS} + \varphi(\alpha)\sin(\alpha_1) \end{cases} \tag{6.20}$$

After the identification of the scatterer positions, the algorithm computes the vector $\tau_s = [\ \tau_R, \quad \tau_{R2}, \quad \cdots \quad \tau_{RN}\ ]$ containing the propagation delays from the MT to the scatterers:

$$\tau_{Ri} = \tau_i - \varphi(\alpha)/c \qquad (i = 1, \ldots, N) \tag{6.21}$$

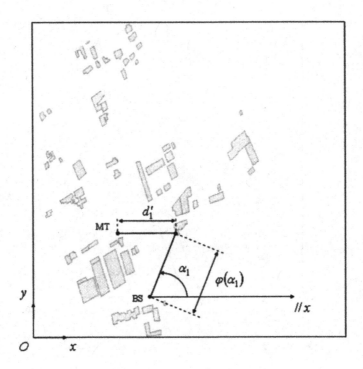

Fig. 6.25 If the mobile terminal MT and the BS are in NLOS, then $\varphi(\alpha_1) < d_1$. Source: M. Porretta, P. Nepa, G. Manara, F. Giannetti, M. Dohler, B. Allen, A. H. Aghvami, *A Novel Single Base Station Location Technique for Microcellular Wireless Networks: Description and Validation by a Deterministic Propagation Model*, IEEE Trans. Veh. Tech., Vol 53, No 5, pp 1553–1560, Sept, 2004 © IEEE.

Then, the following cost function is introduced:

$$F(x, y) = \sum_{i=1}^{N} f_i^2(x, y) \tag{6.22}$$

with

$$f_1(x, y) = c\tau_{Ri} - \sqrt{(x - x_{si})^2 - (y - y_{si})^2} \tag{6.23}$$

Since the MT cannot lie further than $c\tau_1$ from the BS, the estimated position of the MT $(\hat{x}_{MT}, \hat{y}_{MT})$ is eventually chosen as:

$$(\hat{x}_{MT}, \hat{y}_{MT}) = arg \min_{(x,y) \in D} \{F(x, y)\} \tag{6.24}$$

with

$$D = \left\{ (x, y) \mid \sqrt{(x - x_{BS})^2 + (y - y_{BS})^2} \le c\tau_1 \right\} \qquad (6.25)$$

The cost function $F(x, y)$ has been chosen in accordance to the quadratic form $G(x, y)$.

Again, several minimisation techniques can be used to solve the NL-LS problem in equation (6.24) [146]. However, the most straightforward method is by direct comparison. The set of the possible MT locations is sampled at a sufficient step (e.g. 5 m) and the cost function is evaluated for each sample; only the samples belonging to D are considered in the minimisation process and the position $(\hat{x}_{MT}, \hat{y}_{MT})$ is finally chosen which gives the minimum value for $F(x, y)$. Performance degradation can be observed if most of the N MPCs reach the BS after undergoing multiple reflections and/or diffractions. Such a phenomenon, however, can be effectively circumvented by introducing a proper weighting vector $\mathbf{P} = [\ P_1 \quad P_2 \quad \dots \quad P_N\ ]$ in the cost function, thus equation (6.22) becomes

$$F(x, y) = \sum_{i=1}^{N} P_i^2 f_i^2(x, y) \qquad (6.26)$$

in order to account for the probability that the MPC i underwent just a first-order reflection or diffraction. The $\mathbf{P}$ vector can be derived on the basis of heuristic considerations. Indeed, the smaller τ_{Ri} becomes, the more probable is the event that the MPC i underwent just a first-order reflection or diffraction. In the developments of this work, a number of possible functions $P_i = (\tau_{Ri})$ has been considered. Each function $P_i = (\tau_{Ri})$

- satisfies the condition $0 \le P_i(\tau_{Ri}) \le 1$;

- takes its maximum value, which is assumed equal to 1, if $\tau_{Ri} = \tau_{Ri_{min}} = \min_i\{\tau\}$; and

- decreases if τ_{Ri} increases.

Extensive simulations performed in different scenarios demonstrated that a good choice for the weighting vector is given by

$$\begin{cases} 1 & \text{if } \tau_{Ri_{min}} \le \tau_{Ri} \le (1.01 \cdot \tau Ri_{min}) \\ \dfrac{1}{100} \left(\dfrac{\tau_{Ri_{min}}}{\tau_{Ri} - \tau_{Ri_{min}}} \right) & \text{if } \tau_{Ri} \ge (1.10 \cdot \tau_{Ri_{min}}) \end{cases} \qquad (6.27)$$

for $i = 1, \dots, N$.

Nevertheless, a further objective in the investigation is to state an 'optimum' analytical criterion for the choice of vector, $\mathbf{P}$.

6.5.3.2 Numerical results. The performance of the novel location technique described in the previous section has been evaluated in terms of the location error $\varepsilon_d = \sqrt{(x_{MT} - \hat{x}_{MT})^2 + (y_{MT} - \hat{y}_{MT})^2}$. If the LOS condition is verified, then the location $(\hat{x}_{MT}, \hat{y}_{MT})$ is estimated by (6.19), otherwise by (6.24). It is assumed that the actual MT positions are located on a rectangular grid of points and only points lying outside the buildings are taken into account.

The required ToAs and the AoAs of the various MPCs impinging at the BS are calculated through a deterministic propagation model. It is assumed here that the resolution or precision of the ToA and AoA estimation algorithms is infinite, leading to a perfect knowledge of both τ and α. $N = 6$ MPCs impinging on the BS have been assumed for all the numerical results presented throughout this section. However, the sensitivity of the location accuracy with respect to the number, N, of MPCs processed by the algorithm has also been investigated.

Different microcellular scenarios have been considered in the subsequent analysis. For each scenario, the location error, ε_d, and the location error distribution function is analysed. In order to assess the performance of the proposed location technique when the MT is in LOS and in NLOS with the BS, the location error distribution functions conditioned on the events 'MT in LOS with the BS' and 'MT in NLOS with the BS' are also represented, together with the overall location error distribution function.

Finally, simulation results are be compared with US Federal Communication Commission (FCC) results, requiring service providers to be able to locate E-911 callers with an accuracy of 125 m in 67% of cases [147]. FCC requirements are satisfied if the overall location error distribution function is above the point where the vertical line corresponding to 125 m intersects the horizontal line corresponding to 0.67 in all consecutive figures.

The 16-building Manhattan environment is first considered, i.e., a grid of 16 tall buildings. Again, a spacing of 5 m is chosen for the grid of actual MT positions. Such an assumption leads to 3297 actual user positions. Two different positions are chosen for the BS antenna. First, it is located at the centre of a street junction (in the middle of the scenario); then it is mounted on a corner of a building. In both cases, the height of the BS antenna is 5 m above ground level. When the BS antenna is at the centre of the scenario (figure 6.26(a)), a location error with a mean value of 31 m and a standard deviation of 40 m is obtained (figure 6.26(b)). In the asymmetrical situation where the BS antenna is mounted on a corner of a building (figure 6.27(a)), a limited performance degradation is observed. The mean value of the location error is 54 m, while the standard deviation is 68 m (figure 6.27(b)).

Reference [145] compares the performance of this technique to one that requires three BSs to locate the user. The results show a comparable performance to this technique which only requires a single BS to locate the user.

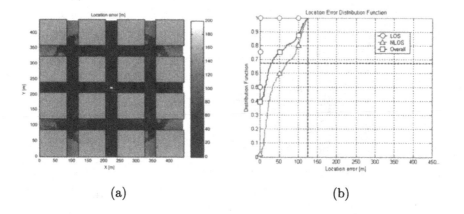

(a)   (b)

Fig. 6.26 Performance of the location technique: 16-building Manhattan environment with BS located at the centre of the scenario. (a) Location error map and (b) location error distribution. The mean value of the location error is 31 m and the standard deviation is 40 m. Source: M. Porretta, P. Nepa, G. Manara, F. Giannetti, M. Dohler, B. Allen, A. H. Aghvami, *A Novel Single Base Station Location Technique for Microcellular Wireless Networks: Description and Validation by a Deterministic Propagation Model*, IEEE Trans. Veh. Tech., Vol 53, No 5, pp 1553–1560, Sept, 2004 © IEEE.

6.5.4 Beamforming for Satellite Communications

6.5.4.1 Current and future satellite systems. In the field of satellite personal communications, multiple beam antennas are required to provide the gain, beam shaping and frequency reuse requirements of such missions. Multiple spot beam antennas can ensure higher *equivalent isotropically radiated power* (EIRP) and G/T (*ratio of the antenna gain to the system noise temperature*) than those achievable by regional coverage shaped beam antennas, while at the same time employing frequency reuse techniques in order to increase the overall system throughput. By using frequency reuse schemes, the same spectrum can be reused multiple times over the coverage area as shown in figure 6.28. However, in practice, frequency reuse is limited by the achievable beam to beam isolation and interference rejection. Therefore, a solution with adaptive arrays may be considered in this case.

Medium earth orbit (MEO) and *low earth orbit* (LEO) satellite constellations require demanding scanning capabilities. With the field of view up to ±60°, these satellites use *directly radiating antenna* (DRA) arrays instead of reflector *multiple beam antenna* (MBA) configurations, as DRAs provide better off-boresight performance with lower scan loss and consequently lower mass. DRA arrays consist of smaller apertures and fewer controlled elements. The employment

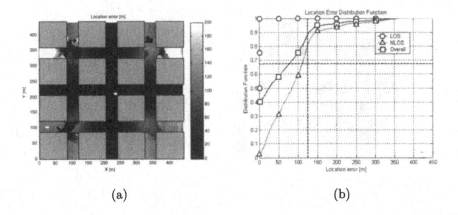

(a)   (b)

Fig. 6.27 Performance of the location technique: 16-building Manhattan environment with BS located on the corner of a building. (a) Location error map and (b) location error distribution. The mean value of the location error is 54 m and the standard deviation is 68 m. Source: M. Porretta, P. Nepa, G. Manara, F. Giannetti, M. Dohler, B. Allen, A. H. Aghvami, *A Novel Single Base Station Location Technique for Microcellular Wireless Networks: Description and Validation by a Deterministic Propagation Model*, IEEE Trans. Veh. Tech., Vol 53, No 5, pp 1553–1560, Sept, 2004 © IEEE.

of an antenna array and the fact that digital processing of the antenna signals is desirable (mainly, because of the large number of spot-beams), provide a strong reason to consider the employment of adaptive antennas onboard these satellites. The advantages of this concept include better suppression of co-channel and inter-channel interference, maximum antenna gain, fixed cells with respect to the user, less handovers, reduced transmit power, higher achievable signal-to-noise-plus-interference ratio and consequently increased system capacity. The main disadvantage is the increased complexity of the hardware since adaptive beamforming will be performed for each user or for a group of users in a common cell. Future satellite communications missions are considered as the means to realise the convergence of broadband (multimedia) applications with mobility. This forms part of the 3G UMTS standard referred to as S-UMTS or satellite UMTS. The two main non-geostationary systems, Globalstar and ICO, together with some regional *geostationary orbit* (GEO) systems (ACeS, Thuraya, EAST), triggered a substantial break-through in satellite active antenna array techniques. Major advances have been achieved in printed array technologies, receive and transmit RF module miniaturisation and performance, beam-forming networks in multilayer printed-circuit technology, digital beam-forming and signal processing techniques, antenna design and integration techniques. Multimedia LEO constellations working at Ka-band (downlink frequency: 20 GHz, uplink

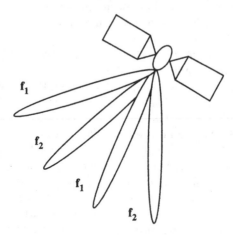

Fig. 6.28 Frequency reuse scheme.

frequency: 30 GHz, bandwidth: 3500 MHz) and Ku-band (downlink frequency: 11 GHz, uplink frequency: 14 GHz, bandwidth: 500 MHz) rely on the development of active multibeam antennas in a direct radiating array configuration. Multimedia space systems based on GEO satellites, such as Euroskyway, are designed to use more traditional multibeam reflector antennas but adaptive approaches could be adopted to achieve traffic reconfigurability and dynamic rain loss compensation in the downlink. Adaptive antenna arrays can also be envisaged for future navigation satellite systems. The possibility of a higher number of radiating elements in the *global positioning system* (GPS) transmitting array is already being considered in order to provide either high EIRP spot beams over the coverage or adaptive nulling capabilities. Similar approaches could be advantageously adopted in the European navigational system called Galileo [153].

It has to be taken into account that the architecture, implementation and performance of an antenna array ultimately determine the communications capacity and the spacecraft's physical characteristics, i.e. configuration, weight and size. Therefore, the active antenna configuration best suited to a specific satellite communications mission has to be selected on the basis of trade-off analyses on many factors, such as EIRP, G/T, mass, power consumption, reliability, overall efficiency and, last but not least, production cost.

The efficiency of future satellite communications and navigation systems can be significantly improved by the use of adaptive antenna arrays with intelligent beamforming and beam-steering. Degradation of the system performance caused by multipath propagation or interferers can be reduced by limited bandwidth, sidelobe reduction, and nulling in the direction of the interferer. This leads to a better quality of service in communication systems and enhances the accuracy of positioning in navigation systems, which play a major role in safety of life services.

6.5.4.2 Switched-beam receiver structure (Butler matrix). As described previously, switched beam antennas are simple in terms of structure and have low cost of production. A linear RF network (called 'Butler matrix' and as depicted in figure 6.16) combines the signals arriving from the L identical antenna elements in order to form different beams looking into certain directions. At the output of a Butler matrix, M RF signals are available. Each of these signals corresponds to a certain beam or direction. Using a simple RF switch, a selection of the best signal takes place followed by further processing by a standard receiver. This signal could be the one with the highest receive power or the best bit error rate (BER). The major advantage of this method is that it requires only a single receive/transmit chain. However, compared to fully adaptive solutions, beams and nulls of the antenna pattern can only be chosen from a limited number of positions and cannot be put into arbitrary positions. Therefore, users that are placed away from the beam peaks or in the centre of a null will suffer from suboptimal services due to the cusping loss, as described earlier in this chapter. The basic switched beam receiver structure is shown in figure 6.29 where the L antenna elements and associated beams are depicted along with the RF switch and subsequent circuitry.

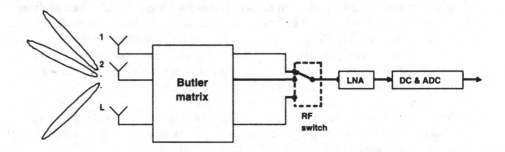

Fig. 6.29 Switched beam receiver structure.

6.5.4.3 Case Study: IRIDIUM. The IRIDIUM satellite system consisted of 66 satellites in six LEO orbits, each at an altitude of 750 km. This system was designed to provide voice and data communications worldwide using handheld terminals in a direct way as shown in figure 6.30. Such a service is similar to cellular telephony with the difference that it is available on a global scale. The key component of each of the IRIDIUM satellites is the main mission antenna, which consists of three fully phased array panels and provides the L-band link from the satellite to the ground user. Each phased array produces 16 fixed simultaneous beams from a total of 48 beams per satellite. The users on the ground communicate with the IRIDIUM network through the satellite whose beam covers the user. As satellites in the network travel along their orbits, the ground user is handed off from beam to beam and from satellite to satellite, depending on which particular beam provides the best link margin to the user. The array beamforming strategy adapted in the IRIDIUM satellite design is based on an RF feed composed of crossed Butler matrices. Different beamlets (i.e. intermediate beams) of the Butler matrices are combined into the desired 16 final beams. The IRIDIUM beamformer consists of eight 16×16 Butler matrices fed in turn by ten 8×8 orthogonal (number of input ports equal to the number of output ports) Butler matrices. The resulting beams of this system fill the scan volume of the array, and meet the IRIDIUM *equivalent isotropic radiated power* (EIRP) and G/T requirements with a margin.

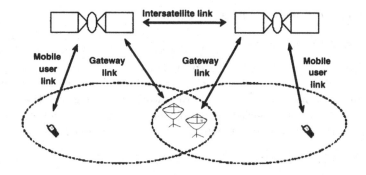

Fig. 6.30   IRIDIUM satellite system.

6.6 CHAPTER SUMMARY

This chapter has presented a number of applications of antenna arrays and associated signal processing techniques. Application areas of:

- biomedical;

- radar;

- sonar;

- terrestrial wireless communications; and

- satellite wireless communications

have been discussed. The illustrations have given a context to many of the techniques introduced throughout the book. It has shown that adaptive antennas can be applied to a wide range of areas and have improved the performance of the retrospective system. The key to the performance gain has been in the choice of array and hardware architecture and associated signal processing techniques. Proper matching of these with the operating scenario and wave propagation characteristics will enable optimum performance gains to be realised.

Although much research has been completed, which emanates largely from the 1950's, there is still a bright future for adaptive antennas. Particular future applications include:

- ultra-wideband communications;

- bio-medical imaging;

- future mobile communications; and

- space applications.

This book has aimed to provide a cross-section of adaptive antenna fundamentals and techniques with a number of applications in mind. Mobile communications has provided a framework which enables the fundamentals to be illustrated, but attention has also been given to radar and bio-medical systems. Many of the principles and techniques are transferable to applications not discussed here.

6.7 PROBLEMS

Problem 1. List four applications of adaptive antenna systems.

Problem 2. What is the main advantage of replacing mechanically steered antennas with electrically steerable antennas?

Problem 3. Describe why thermal management is important in medical ultra-sonic array design.

Problem 4. Describe two geometries to focus short axis elevation in medical ultra-sonic arrays.

Problem 5. List the parameters that affect image quality from medical ultra-sonic arrays.

Problem 6. Describe three losses that occur in switched beam beamforming systems.

Problem 7. Explain why adaptive antenna arrays could improve the performance of satellite communication systems.   In which ways is the link efficiency of such systems expected to be affected?

Problem 8. Discuss the differences between a switched beam receiver (Butler matrix) and an adaptive receiver employed by an antenna array.

References

1. P. W. Howells. Intermediate frequency sidelobe canceller. *Technical report, U.S. Patent 3202990*, May 1959.

2. S. P. Applebaum. Adaptive arrays. *IEEE Trans. Antennas and Propagation*, 24:585–598, September 1976.

3. B. Widrow, P. E. Mantey, L. J. Griffiths, and B. B. Goode. Adaptive antenna systems. *Proceedings of the IEEE*, 55(12):2143–2159, December 1967.

4. L. J. Griffiths. A simple adaptive algorithm for real-time processing in antenna arrays. *IEEE Proceedings*, 57:1696–1704, October 1969.

5. O. L. Frost. An algorithm for linearly constrained adaptive array processing. *IEEE Proceedings*, 60:926–935, August 1972.

6. R. Compton, R. Huff, W. Swarner, and A. Ksienski. Adaptive arrays for communication systems: An overview of research at the Ohio State University. *IEEE Trans Antennas and Propagation*, 60(5):599–607, September 1976.

7. B. Allen and M. Beach. On the analysis of switch-beam antennas for the w-cdma downlink. *IEEE Transactions on Vehicle Technology*, 53(3):569–578, May 2004.

8. M. Ghavami, L. B. Michael, and R. Kohno. *Ultra Wideband Signals and Systems in Communications Engineering.* John Wiley & Sons, Ltd, May 2004.

9. C. Balanis. *Antenna Theory, Analysis and Design,* page 46. John Wiley & Sons, Ltd, 1982.

10. R. Munson. Conformal microstrip antennas and microstrip phased arrays. *IEEE Transactions on Antennas and Propagation,* 22(1):74–78, January 1975.

11. H. Assumpcao and G. Moutford. An overview of signal processing for arrays and receivers. *Journal Int. Eng. Aus.t and IREE Aust.,* 4:6–19, 1984.

12. V. Anderson. Dicanne, a realizable adaptive process. *Journal Acoust. Soc. Amer.,* 45:398–405, 1969.

13. K. Hugl, L. Laurila, and E. Bonek. Downlink performance of adaptive antennas with null broadening. *IEEE Vehicular Technology Conference Proceedings,* pages 872–876, 1999.

14. M. Hunukumbure, M. Beach, B. Allen, P. Fletcher, and P. Karlsson. Smart antenna performance degradation due to grating lobes in fdd systems. *Proceedings of 9th COST 260 Meeting on Smart Antennas: CAD and Technology,* May 2001.

15. M. Skolnik. *Introduction to Radar Systems,* page 284, 1980.

16. S.Deghan, D. Lister, R. Owen, and P. Jones. W-cdma capacity and planning issues. *IEE Electronics and Communications Engineering,* 12(3):101–118, June 2000.

17. A. V. Oppenheim and R. W. Schafer. *Discrete Time Signal Processing,* pages 447–448. Prentice Hall, 1989.

18. J. F. Kaiser. Nonreeursive digital filter design using the io-sinh window function. *IEEE Circuits and Systems Conference,* pages 20–23, April 1974.

19. M. Ghavami. Wide-band smart antenna theory using rectangular array structures. *IEEE Transactions on Signal Processing,* 50(9):2143–2151, September 2002.

20. D. B. Ward, R. A. Kennedy, and R. C. Williamson. Fir filter design for frequency invariant beamformers. *IEEE Signal Processing Letters,* 3(3):69–71, March 1996.

21. D. B. Ward, R. A. Kennedy, and R. C. Williamson. Theory and design of broadband sensor arrays with frequency invariant far field beam patterns. *Journal of Acoustic Society of America*, 97(2):1023–1034, February 1995.

22. G. Tsoulos. Smart antennas for third generation wireless personal communications. *University of Bristol PhD Thesis*, pages 55–59, 1997.

23. T. MacNamara. Simplified design procedure for butler matrices incorporating 90 degree hybrids. *IEE Proceedings Part H*, 134(1):50–54, January 1987.

24. R. Hansen. *Microwave Scanning Arrays*, volume 3, pages 258–263. Academic Press, 1966.

25. J. Tui. *Digital Microwave Receivers, Theory and Concepts*, pages 250–253. Artech House, 1989.

26. D. Archer. Lens fed multiple beam arrays. *Microwave Journal*, pages 171–195, September 1984.

27. P. Howard, C. Simmonds, P. Darwood, M. Beach, R. Arnot, and F. Cesbron. Adaptive antenna performance in a mobile system. *TSUNAMI II Technical Report AC020/ORA/WP3/DS/P/008/a1*, 1998.

28. P. Mogensen, P. Zetterberg, H. Dam, P. Espensen, and F. Fredriksen. Algorithms and antenna array recommendations. *TSUNAMI II Technical Report AC020/AUC/A1.2/DR/P/005/b1*, 1997.

29. T. Moorti and A. Paulraj. Performance of switched beam systems in cellular base stations. *IEEE ASILOMAR 29 Conference Proceedings*, pages 388–392, 1995.

30. T. Bachelier, J. Sante, and G. Cegetel. Influence of mobility on capacity of dcs network using switch beam antenna. *Intelligent Antenna Technology for Mobile Communications Symposium*, 1997.

31. C. Ward, D. Adams, F. Wilson, and A. Bush. The live-air trial of a multi-beam cellular base station antenna system. *IEE National Conference on Antennas and Propagation*, pages 169–172, 1999.

32. M. Ho, G. Stuber, and M. Austin. Performance of switch beam smart antennas for cellular radio systems. *IEEE Transactions on Vehicular Technology*, 47(1):10–19, February 1998.

33. L. Godara. Application of antenna arrays to mobile communications, part ii: Beamforming and direction of arrival considerations. *Proceedings of the IEEE*, 85(8):1195–1245, August 1997.

34. S. Haykin. *Adaptive Filter Theory*. Prentice Hall, pages 165–168, 1996.

35. J. Litva and T. Lo. *Digital Beamforming in Wireless Communications*, pages (a): 13–34, (b): 57–91, (c) 13–27. Artech House, 1996.

36. R. Piechocki, N. Canagarajah, J. McGeehan, and G. Tsoulos. Orthogonal re-spread for uplink wcdma beamforming. *IEEE Vehicular Technology Conference*, May 2000.

37. M. Bengtsson and B. Ottersten. Optimum downlink beamforming using semi-definite optimisation. *37th Annual Allerton Conference on Communication, Control and Computing*, June 1999.

38. H. Krim and M. Viberg. Two decades of array signal processing. *IEEE Signal Processing Magazine*, July 1996.

39. M. Bartlett. Smoothing periodograms from time series with continuous spectra. *Nature*, (161), 1948.

40. J. Capon. High resolution frequency-wavenumber spectrum analysis. *IEEE Proceedings*, 57(8), August 1969.

41. R. Schmidt. A subspace approach to multiple emitter location and spectral estimation. *PhD Thesis, University of Stanford*, 1981.

42. J. Bohme. Estimation of source parameters by maximum likelihood and non-linear regression. *IEEE Conference of Acoustics, Speech and Signal Processing*, pages 731–734, 1984.

43. A. Swindlehurst and P. Stoica. Maximum likelihood methods in radar array signal processing. *IEEE Proceedings*, 86(2):421–441, 1998.

44. A. Paulraj, R. Roy, and T. Kailath. A subspace rotation approach to signal parameter estimation. *IEEE Proceedings*, 74(7):1044–1045, July 1986.

45. J. Fessler and H. Hero. Space alternating generalised expectation maximisation algorithm. *IEEE Transactions on Signal Processing*, 42(10):2664–2677, October 1994.

46. M. Viberg, B. Ottersten, and T. Kailath. Detection and estimation in sensor arrays based on weighted subspace fitting. *IEEE Transactions on Signal Processing*, 39(11):2436–22449, November 1991.

47. M. Haardt and J. Nossek. Unitary esprit: How to obtain increased estimation accuracy with reduced computational burden. *IEEE Transactions on Signal Processing*, 43(5):1232–1242, May 1995.

48. B. Widrow and S. D Stearns. *Adaptive Signal Processing*. Prentice Hall, 1985.

49. T. Rappaport. *Wireless Communications: Principles and Practice.* Prentice Hall, 1996.

50. R. Screiber. Implementation of adaptive array algorithms. *IEEE Trans. Acoust., Speech, Signal Processing,* pages 1038–1045, 1986.

51. B. D. VanVeen. Adaptive convergence of linearly constrained beamformers based on the sample covariance matrix. *IEEE Trans. Signal Processing,* 39:1470–1473, 1991.

52. J. H. Winters, J. Salz, and R. D. Gitlin. The impact of antenna diversity on the capacity of wireless communication systems. *IEEE Trans. Comm.,* 42:1740–1751, 1994.

53. T. Gebauer and H. G. Gockler. Channel-individual adaptive beamforming for mobile satellite communications. *IEEE Trans. Comm.,* 13:439–448, 1995.

54. R. G. Vaughan. On optimum combining at the mobile. *IEEE Trans. Veh. Tech.,* 37:181–188, 1988.

55. C. Passerini, M. Missiroli, G. Riva, and M. Frullone. Adaptive antenna arrays for reducing the delay spread in indoor radio channels. *IEE Electronics Letters,* 32:280–281, 1996.

56. J.Capon. High resolution frequency-wavenumber spectral analysis. *Proc. Of the IEEE,* 57(8):1408–1418, August 1969.

57. R.O. Schmidt. Multiple emitter location and signal parameter estimation. *IEEE Trans. On Antennas and Propagation,* 34(3), March 1986.

58. A. J. Barabell. Improving the resolution performance of eigenstructure-based direction finding algorithms. *Proc. of the IEEE Int'l. Conf. on Acoustics, Speech, and Signal Processing,* pages 336–339, 1983.

59. S. V. Schell, Calabretta, W. A. Gardner, and B. G. Agee. Cyclic music algorithms for signal selective doa estimation. *Proc. of the IEEE Int'l. Conf. on Acoustics, Speech, and Signal Processing,* pages 2278–2281, 1989.

60. G. Xu and T. Kaiath. Fast subspace decomposition. *IEEE Trans. on Signal Processing,* 42(3):539–550, March 1994.

61. R. S. Thoma, D. Hampicke, A. Richter, G. Sommerkorn, A. Schneider, U. Trautwein, and W. Wirnitzer. Identification of time-variant directional mobile radio channels. *IEEE Transactions on Instrumentation and Measurement,* 49(2):357–364, April 2000.

62. M. Haardt and J. Nossek. Structured least squares to improve the performance of esprit-type algorithms. *IEEE Transactions on Signal Processing*, 45(5), May 1997.

63. A. Richter, D. Hanpicke, G. Sommerkorn, and R. Tmoma. Joint estimation of toa, time-delay and doa for high-resolution channel sounding. *IEEE Vehicle Technology Conference*, 2000.

64. I. Ziskind and M. Wax. Maximum likelihood localization of multiple sources by alternating projection. *IEEE Trans. ASSP*, 36(10):1553–1560, October 1988.

65. J. A. Fessler and A. O. Hero. Space-alternating generalised expectation maximisation algorithm. *IEEE Trans. Signal Processing*, 42:2664–2677, October 1994.

66. J. E. Evans, J. R. Johnson, and D. F. Sun. High resolution angular spectrum estimation techniques for terrain scattering analysis and angle of arrival estimation in atc navigation and surveillance system. *M.I.T. Lincoln Lab., Lexington, MA, Rep.*, page 582, 1982.

67. H. Akaike. A new look at statistical model identification. *IEEE Trans. on Automatic Control*, 19:716–723, 1974.

68. J. Rissanen. A universal prior for integers and estimation by minimum description length. *Ann. of Statistics*, 11:416–431, 1983.

69. J. Laurila. *Semi-blind Detection of Co-channel Signals in Mobile Communications*. PhD Thesis, Technische Universitat Wien, 2000.

70. B. G. Agee, S. V. Schell, and W. A. Gardner. Spectral self-coherence restoral: a new approach to blind adaptive signal extraction using antenna arrays. *IEEE Proceedings*, 78(4):753–767, 1990.

71. T. E. Biedka. A method for reducing computations in cyclostationarity-exploiting beamforming. *IEEE ICASSP*, pages 1828–1831, 1995.

72. K. Nishimori, N. Kikuma, and N. Inagaki. The differential cma adaptive array antenna using an eigen-beamspace system. *IEICE Trans. on Communications*, (11):1480–1488, November 1995.

73. Z. Rong, P. Petrus, T. S. Rappaport, and J. H. Reed. Despread-respread multi-target constant modulus array for cdma systems. *IEEE Communications Letters*, 1(4):114–116, July 1997.

74. Non-penetrating mast mount 6-1/2ft x 6-1/2ft. *Prodelin Corporation*, April 2002.

75. Andrew. http://www.andrew.com.

76. B. Sklar.   *Digital Communications*,   pages 213–225.   Prentice-Hall International Inc, 1988.

77. J. D. Fredrick, Y. Wang, and T. Itoh.   Smart antennas based on spatial multiplexing of local elements (smile) for mutual coupling reduction. *IEEE Transactions on Antennas and Propagation*, 52(1):106–114, January 2004.

78. G. Tsoulos, M. Beach, and J. McGeehan. Wireless personal communications for the 21st century: European technological advances in adaptive antennas. *IEEE Communications Magazine*, 35(9):102–109, September 1997.

79. G. Tsoulos, J. McGeehan, and M. Beach.   Space division multiple access (sdma) field trials. Part2: Calibration and linearity issues. *IEE Proc. radar sonar and navigation*, 145(1):79–84, February 1998.

80. M. E. Frerking. *Digital Signal Processing in Communication Systems*, pages (a): 577–580, (b): 364–367, (c): 353–358. Van Nostrand Reinhold, 1994.

81. WP5 TSUNAMI(2) ACTS AC020. Cost effectiveness of adaptive antennas, 1998.

82. J. C. Liberti and T. S. Rappaport.   *Smart Antennas for Wireless Communications*, pages 91–105. Prentice-Hall PTR, 1999.

83. N. Tyler, B. Allen, and A. H. Aghvami. Adaptive antennas - the calibration problem. *IEEE Communications Magazine*, 42(10), 2004.

84. Mini-circuits. RF/IF designers guide DG-Y2K, p. 168, 2000.

85. R. J. Mailloux. *Phased Array Antenna Handbook*, pages 63–72. Artech House, Inc., 1994.

86. K. R. Dendekar, H. Ling, and G. Xu. Experimental study of mutual coupling compensation in smart antenna applications. *IEEE Transactions on Wireless Communications*, 1(3):480–487, July 2002.

87. K. R. Dendekar, H. Ling, and G. Xu. Effect of mutual coupling on direction finding in smart antenna applications. *Electronic Letters*, 36(22):1889–1891, October 2000.

88. R. O. Schmidt. Multiple emitter location and signal parameter estimation. *IEEE Transactions on Antennas and Propagation*, 34(3):276–280, March 1986.

89. T. Su, K. R. Dendekar, and Ling.   Simulation of mutual coupling effect in circular arrays for direction-finding applications. *Microwave and Optical Technology Letters*, 26(5), 2000.

90. M. K. Ozdemir, H. Arslan, and E. Arvas. A mutual coupling model for MIMO systems. In *IEEE Topical Conference on Wireless Communication Technology*, October 2003.

91. B. K. Lau. Applications of adaptive antennas in third-generation mobile communication systems, PhD thesis, Curtin University of Technology, November 2002.

92. S. J. Orfanidis. Electromagnetic waves & antennas, www.ece.rutgers.edu/~orfanidi/ewa, pp. 664-688, February 2004.

93. A. CZylwik and A. Dekorsy. System level simulations for downlink beamforming with different array topologies. In *IEEE GLOBECOM*, November 2001.

94. W. Weichselberger and G. Lopez de Echazarreta. Comparison of circular and linear antenna arrays with respect to the UMTS link level. In *COST 260 meeting and workshop, Gothenburg, Sweden*, May 2001.

95. B. K. Lau and Y. H. Leung. A Dolph-Chebyshev approach to the synthesis of array patterns for uniform circular arrays. In *IEEE International Symposium on Circuits and Systems*, May 2000.

96. B. K. Lau and Y. H. Leung. Optimum beamformers for uniform circular arrays in a correlated signal environment. In *IEEE International Conference on Acoustics, Speech and Signal Processing*, June 2000.

97. B. K. Lau, Y. H. Leung, Y. Liu, and K. L. Teo. Direction of arrival estimation in the presence of correlated signals and array imperfections with uniform circular arrays. In *IEEE International Conference on Acoustics, Speech and Signals Processing*, May 2002.

98. H. R. Anderson and J. P. McGeehan. Direct calculation of coherence bandwidth in urban micro-cells using ray-tracing propagation model. *IEEE Personal Indoor and Mobile Communications Conference*, pages 20–24, 1994.

99. B. Allen, J. Webber, P. Karlsson, and M. Beach. UMTS spatio-temporal propagation trail results. *IEE International Conference on Antennas and Propagation*, April 2001.

100. J. K. Cavers. *Mobile Channel Characteristics*, pages 1–113. Kluwer Academic Publishers, 2000.

101. S. R. Saunders. *Antennas and Propagation for Wireless Communications Systems*, pages 85–321. John Wiley & Sons, Ltd, 2000.

102. R. Ertel, P. Cardieri, K. Sowerby, T. Rappaport, and J. Reed. Overview of spatial channel models for antenna array communication systems. *IEEE Personal Communications Magazine*, 5(1):10–22, February 1998.

103. M. Schwartz. *Information Transmission, Modulation and Noise*, pages 579–582. McGraw-Hill, 1990.

104. B. Allen, M. Beach, and P. Karlsson. Analysis of smart antenna outage in utra fdd networks. *IEE Electronics Letters*, 38(1):2–3, 2002.

105. Physical channels and mapping of transport channels onto transport channels (tdd). *3GPP Specification Doc Number TS25221 v3.0.0*, October 1999.

106. A. Paulraj and C. Papadias. Space-time processing for wireless communications. *IEEE Signal Processing Magazine*, November 1997.

107. www.gsmworld.com/technology/spectrum_gsm.html.

108. Ue radio transmission and reception (fdd). *3GPP Specification Doc Number TS25.101 v3.1.0*, December 1999.

109. K. Hugl and E. Bonek. Performance of downlink nulling for combined packet/circuit switched systems. *IEEE Vehicle Technology Conference*, September 2000.

110. K. Gilhousen, I. Jacobs, R. Padovani, A. Viterbi, L. Weaver, and C. Wheatley. On the capacity of a cellular cdma system. *IEEE Transactions on Vehicle Technology*, 40(2):303–311, May 1991.

111. M. A. Beach, C. M. Simmonds, P. Howard, and P. Darwood. European smart antenna test-bed - field trial results. *IEICE Trans. on Communications*, (9):2348–2356, September 2001.

112. Physical layer procedures. *3GPP Specification Doc Number TS25.214 v3.1.1*, December 1999.

113. http://www.medav.de.

114. P. Zetterberg and B. Ottersen. The spectral efficiency of a base station antenna array system for spatially selective transmissions. *IEEE Transactions on Vehicular Technology*, 44(3):651–660, August 1995.

115. B. Lindmark, M. Ahlberg, M. Nilsson, and C. Beckman. Performance analysis of applying uplink estimates in the downlink beamforming using a dual polarised array. *IEEE Vehicular Technology Conference Proceedings*, 2000.

116. J. Thompson, P. Grant, and B. Mulgrew. Performance of downlink beamforming techniques for cdma. *Intelligent Antenna Technology for Mobile Communications Symposium*, 1998.

117. M. Kitahara, Y. Ogawa, and T. Ohgane. A base station adaptive antenna for downlink transmissions in a ds-cdma system. *IEEE Vehicular Technology Conference Proceedings*, 2000.

118. K. Hugl, L. Laurila, and E. Bonek. Downlink beamforming for frequency division duplex systems. *IEEE Globecom Conference Proceedings*, pages 2097–2101, 1999.

119. K. Hugl. Frequency transformation based downlink beamforming. *COST 259/260 Joint Workshop Proceedings*, pages 111–118, April 1999.

120. J. Goldberg and J. Fonollosa. Downlink beamforming for cellular mobile communications. *IEEE Vehicular Technology Conference*, pages 632–636, 1997.

121. G. Rayleigh, S. Diggavi, V. Jones, and A. Paulraj. A blind adaptive transmit antenna algorithm for wireless communications. *IEEE International Communications Conference Proceedings*, pages 1494–1499, 1995.

122. K. Hugl, L. Laurila, and E. Bonek. Smart antenna downlink beamforming for uncorrelated communications links. *AP2000 Conference Proceedings*, pages 2097–2101, 2000.

123. W. Utschick and J. Nossek. Downlink beamforming for fdd mobile radio systems based on spatial covariances. *European Wireless 99 and ITG Mobile Communications Conference Proceedings*, pages 65–67, 1999.

124. T. Aste, P. Forester, L. Fety, and S. Mayrargue. Downlink beamforming avoiding doa estimation for cellular mobile communications. *IEEE International Conference on Acoustics, Speech and Signal Processing*, 1998.

125. D. Gerlach and A. Paulraj. Base station transmitting antenna arrays for multipath environments. *EASP Signal Processing Journal*, 54(1):59–73, October 1996.

126. D. Gerlach and A. Paulraj. Adaptine transmitting antenna arrays with feedback. *IEEE Signal Processing Letters*, 1(10):150–152, October 1994.

127. D. Gerlach and A. Paulraj. Adaptive transmitting antenna methods for multipath environments. *IEEE Globecom Conference Proceedings*, pages 425–429, 1994.

128. J. Choi, S. Perreau, and Y. Lee. Semi-blind method for transmit antenna array in cdma systems. *IEEE Vehicular Technology Conference Proceedings*, 2000.

129. J. Thompson, J. Hudson, P. Grant, and B. Mulgrew. Cdma downlink beamforming for frequency selective channels. *IEEE Personal Indoor Mobile Radio Conference Proceedings*, 1999.

130. G. Auer, J. Thompson, and P. Grant. Performance of antenna array transmission techniques for cdma. *IEE Electronics Letters*, 33(5):369–370, 1997.

131. E. Tirrola and J. Ylitalo. Performance of fixed-beam beamforming in wcdma downlink. *IEEE Vehicular Technology Conference Proceedings*, 2000.

132. T. Mootri, R. Stutzle, and A. Paulraj. Performance of a fixed-beam system in the IS-95 cdma forward link. *European Transactions on Telecommunications*, 9(4):361–370, 1998.

133. G. Rayleigh and V. Jones. Adaptive transmission for frequency division duplex digital wireless communications. *IEEE International Communications Conference*, pages 641–646, June 1997.

134. F. Rashid-Farrokhi. Transmit beamforming and power control for cellular wireless communications. *IEEE Journal on Selected Areas of Communications*, 16(8), October 1998.

135. M. Bengtsson. Optimum transmission using smart antennas. *IST Mobile Communications Summit*, pages 359–364, June 2000.

136. X. Mestre and J. Fonollosa. Performance evaluation of uplink and downlink beamforming for utra fdd. *Proceedings of FRAMES workshop*, 1999.

137. R. E. McKeighen. Design guidelines for medical ultrasonic arrays. In *SPIE International Symposium on Medical Imaging*, February 1998.

138. E. J. Bond, X. Li, S. C. Hagness, and B. D. Van Veen. Microwave imaging via space-time beamforming for early detection of breast cancer. *IEEE Transactions on Antennas and Propagation*, 51(8):1690–1705, August 2003.

139. M. A. Beach, J. P. McGeehan, C. M. Simmonds, P. Howard, P. Darwood, G. V. Tsoulos, A. R. Nix, P. Hafezi, and Y. Sun. European smart antenna test-beds. *Journal of Communication Networks*, pages 317–324, December 2000.

140. C. M. Simmonds, P. B. Darwood, M. A. Beach, and P. Howard. Tsunami (ii) macrocellular field trial system performance in the presence of deliberate interference. *IEE International Conference on Antennas and Propagation*, pages 73–76, 1999.

141. E. Dahlman, B. Gudmundson, M. Milsson, and J. Skold. UMTS/IMT2000 based on wideband cdma. *IEEE Communications Magazine*, pages 70–80, September 1998.

142. F. Adachi. The effects of orthogonal spreading and rake combining on ds-cdma forward link in mobile radio. *IEICE Transactions on Communications*, (11):1703–1712, 1997.

143. M. Hunukumbure, M. Beach, and B. Allen. On the down-link orthogonality factor in utra fdd systems. *IEE Electronics Letters*, 38(4):196–197, 2002.

144. T. Ojanpera and R. Prasad. *Wideband CDMA for Third Generation Mobile Communications*, pages 226–229. Artech House, 1998.

145. M. Porretta, P. Nepa, G. Manara, F. Giannetti, M. Dohler, B. Allen, and A. H. Aghvami. A novel single base station location technique for microcellular wireless networks: Description and validation by a deterministic propagation model. *IEEE Trans. Vehicular Technology*, 53(5), September 2004.

146. J. J. Caffery and G. L. Stber. Overview of radiolocation in cdma systems. *IEEE Communications Magazine*, 36(4):38–45, April 1998.

147. Revision of the commissions rules to ensure compatibility with enhanced 911 emergency calling systems. *FCC*, RM-8143(CC Docket 94-102), July 1996.

148. M. Hata and T. Nagatsu. Mobile location using signal strength measurements in a cellular system. *IEEE Trans. Veh. Technol.*, 29, May 1980.

149. S. Sakagami et. al. Vehicle position estimates by multi-beam antennas in multipath environments. *IEEE Trans. Veh. Technol.*, 41, February 1992.

150. H. Hashemi. Pulse ranging radiolocation technique and its application in channel assignment in digital cellular radio. *IEEE Veh. Technol. Conf.*, pages 675–680, 1991.

151. D. Kothris, M. Beach, B. Allen, and P. Karlsson. Performance assessment of terrestrial and satellite based position location systems. *IEE Conf. 3G Mobile Communication Technologies*, pages 211–215, 2001.

152. N. J. Thomas, D. G. M. Cruickshank, and D. I. Laurenson. Calculation of mobile location using scatterer information. *IEE Electronics Letters*, 37(19):1193–1194, 2001.

153. C. Forrester. Threat or opportunity? *IEE Communications Engineer Magazine*, pages 10–14, October 2003.

Index